COGNITIVE RADIO

Edited by **Tõnu Trump**

Cognitive Radio
http://dx.doi.org/10.5772/65146
Edited by Tõnu Trump

Contributors

Mustafa Namdar, Arif Basgumus, Yasir Ahmed Al-Mathehaji, Said Boussakta, Martin Johnston, Yingsong Li, Yanyan Wang, Jose R. Garcia Oya, Fernando Muñoz Chavero, Ruben Martin-Clemente, Tonu Trump

Published by InTech
Janeza Trdine 9, 51000 Rijeka, Croatia

Publishing Process Manager Romina Rovan
Technical Editor SPi Global
Cover InTech Design team

First published July, 2017
Printed in Croatia
Legal deposit, Croatia: National and University Library in Zagreb

Additional hard copies can be obtained from orders@intechopen.com

Cognitive Radio, Edited by Tõnu Trump
p. cm.
Print ISBN 978-953-51-3337-7
Online ISBN 978-953-51-3338-4

For EU product safety concerns:
IN TECH d.o.o., Prolaz Marije Krucifikse Kozulić 3, 51000 Rijeka, Croatia,
info@intechopen.com or visit our website at intechopen.com.

Meet the editor

Tõnu Trump received his PhD degree from the Tallinn University of Technology in 1994. After his postgraduate studies at Royal Institute of Technology, Stockholm, he joined Ericsson AB in Stockholm, where he worked until 2006. His final position was as expert in echo cancellation and voice enhancement devices. From 2002 to 2006, he was the rapporteur of study group 16 question 17 'voice gateway equipment' at ITU-T in Geneva. From 2007 to the beginning of 2017 he was professor of signal processing at Tallinn University of Technology. At the present time, he is a consultant at Virgostell OÜ. Prof. Trump has published more than 70 papers and is the author of 11 patents. He is a senior member of IEEE.

Contents

Preface

The subject of cognitive radio is a relatively new area of research. It was introduced in 1998 by Joseph Mitola III and has since then attracted many researches around the word. Cognitive radio stems from the concept of software-defined radio and is an intelligent adaptive radio that monitors its own performance continuously. It determines the radio environment and adjusts its settings to provide the required quality of service. Most importantly, the cognitive radio is able to find the spectrum necessary for its work by itself. In this book, we present several new research papers about cognitive radio written by people of around the word.

In the chapter 'Reliable Broadcast over Cognitive Radio Networks: A Bipartite Graph-Based Algorithm', Y. A. Al-Mathehaji, S. Boussakta and M. Johnston address the problem of reliable broadcast in cognitive radio networks. The focus is on developing a cognitive system that does not use a common control channel to enable control message exchange. They use graph theory, more precisely bipartite graphs, to map the problem at hand to the problem of set cover. Then, they develop an algorithm, which guarantees a distributed reliable selection of the broadcast channel with a facilitative channel switching facility where primary user activity is detected without the need for frequent reselection.

The chapter by M. Namdar and A. Basgumus 'Outage Performance Analysis of Underlay Cognitive Radio Networks with Decode-and-Forward Relaying' evaluates outage performance of relaying over Rayleigh fading channels subject to the relay location for a secondary user. The authors provide optimal location of the relay terminals in cognitive networks. The analysis provides maximum transmission rates of the secondary user and the outage probabilities.

The chapter by J. R. G. Oya, M. Chavero and R. M. Clemente 'Analog-to-Digital Conversion for Cognitive Radio: Subsampling, Interleaving, and Compressive Sensing' focuses on analog-to-digital conversion techniques. The conversion is necessary if the source signal is in analog form like speech or music. The chapter gives an overview of promising techniques like subsampling, interleaving and compressive sensing and discusses solutions of integrating these techniques into a unique analog-to-digital conversion process.

The chapter 'Reconfigurable Antennas for UWB Cognitive Radio Communication Applications' by Y. Li and Y. Wang concentrates on antennas for ultrawideband cognitive radio communication applications. The chapter discusses the defected microstrip structure and uses the structure to form antennas that can filter out interfering narrowband signals. After that, the authors focus on creating notches in the antenna response.

It is hoped that the book is useful for researchers and engineers in both academia and industry working on problems related to radio communications in general and cognitive radio in particular.

Tõnu Trump
Virgostell OÜ,
Tallinn, Estonia

Section 1

Introduction

Chapter 1

Introductory Chapter

Tõnu Trump

Additional information is available at the end of the chapter

http://dx.doi.org/10.5772/intechopen.69597

1. Cognitive radio challenge

One of the most critical resources required for wireless communication is the radio spectrum. One can see the radio spectrum as nonrenewable natural resource. If a part of spectrum has been used for some application, one cannot simultaneously reuse it at the same place for some other application. National and international laws and agreements regulate spectrum usage so that the services provided were free from interference caused by other users. Traditionally, the administration of the spectrum rights tends to grant exclusive rights to some services in the major geographic regions. For instance, in the United States and many other countries, the frequency band 535–1605 kHz is allocated for AM radio, 54–72 MHz for TV channels 2–4, 88–108 MHz for FM radio, and so on. This static allocation has over the years led to many successful applications, but it has also resulted in a situation where almost all the available spectrum has been assigned to specific applications and there is no room for emerging services.

On the other hand, several studies and measurement campaigns are showing that the spectrum is actually underutilized. Spectrum utilization depends on frequency, geographical location and time. Fixed spectrum allocation, however, prevents the rarely used frequency bands being reused. These studies suggest that new devices should use the underutilized spectrum in an opportunistic manner. It leads to the core idea behind cognitive radio, i.e., the radio that is aware of the environment and can adapt the transmissions according to the interference it sees. In other words, the cognitive radio seeks the unutilized frequencies and uses them for its own transmissions in an adaptive manner. The concept of cognitive radio was first proposed by Joseph Mitola III in a seminar at the Royal Institute of Technology in Stockholm in 1998 and published in an article by Mitola and Maguire in 1999 [1]. Since then, there has been a lot of work on the concept some of which is printed in this volume.

Cognitive radio is an adaptive, intelligent radio, and network technology that can automatically detect available channels in a wireless spectrum and change transmission parameters, enabling more communications to be performed concurrently. Cognitive radio is based on software radio technology where the pieces of software have replaced traditional hardware components such as amplifiers, modulators, and mixers. This way it is easy to change the operation of the radio. All that is needed is reprogramming. It can also be considered to be an adaptive radio, which monitors and modifies its own performance.

The cognitive radio needs to collect cognition about the radio environment to operate efficiently. Such a radio needs to understand if the spectrum it intends to use is free or utilized by some primary user and redistribute the available spectrum dynamically. By primary user, we mean the licensed user of the band, and correspondingly, the cognitive radios are often termed as secondary users. This process is called spectrum sensing.

A secondary user may collect information about primary user activities alone or it may cooperate with other secondary users to improve the detection and estimation results. Of course for the cooperation to be possible, several secondary users must be designed so that they allow it. If cooperation between the secondary users is possible, it results in more reliable detection. This is because cooperation allows overbridging the fading and shadowing effects that are present in real-world radio propagation by the usage of spatial diversity, which in turn improves the results.

The first standard on cognitive radio was developed by IEEE (IEEE 802.22) and published in 2011. The standard combines a database of licensed users of the area with spectrum sensing to locate the primary users. The standard was developed for usage of unused television channels in the rural areas.

The chapters of this book discuss different aspects of cognitive radio, covering a large span of the problems that have to be solved in order to build reliable systems. This is the hope of the editor that the material published in this book is useful for people who design the cognitive radio systems and for the people who research the different aspects of the exiting subject.

Author details

Tõnu Trump

Address all correspondence to: tonu.trump@gmail.com

Virgostell OÜ, Tallinn, Estonia

References

[1] Mitola III J, Maguire GQ, Jr. Cognitive radio: Making software radios more personal. IEEE Personal Communications Magazine. 1999;**6**(4):13-18

Section 2

System Aspects

Reliable Broadcast over Cognitive Radio Networks: A Bipartite Graph-Based Algorithm

Yasir Ahmed Al-Mathehaji, Said Boussakta and
Martin Johnston

Additional information is available at the end of the chapter

http://dx.doi.org/10.5772/intechopen.69216

Abstract

Cognitive radio (CR) is a promising technology that aims to enhance the spectrum utilisation by enabling unlicenced users to opportunistically use the vacant spectrum bands assigned to licenced users. Broadcasting is considered as a fundamental operation in wireless networks, as well as in cognitive radio networks (CRNs). The operation of most network protocols in the ad hoc network depends on broadcasting control information from neighbouring nodes. In traditional single-channel or multichannel ad hoc networks, due to uniform channel availability, broadcasting is easily implemented as nodes are tuned to a single common channel. On the contrary, broadcasting in CR ad hoc networks is both a challenging and complex task. The complexity emerges from the fact that different CR users might acquire different channels at different times. Consequently, this partitions the network into different clusters. In this chapter, the problem of broadcasting in ad hoc CR networks is presented, current solutions for the problem are discussed and an intelligent solution for broadcasting based on graph theory to connect different local topologies is developed.

Keywords: reliable broadcast, cognitive radio, bipartite graph, network topology, data dissemination, ad hoc network

1. Introduction

The idea of an intelligent wireless communication framework cognitive radio (CR) network has been proposed as a solution to deal with the disparity between the increasing demand of wireless radio spectrum and the spectrum underutilisation by licenced users [1]. Unlike conventional spectrum policy in which designated parts of the spectrum are allocated specifically for exclusive use to licenced users (usually referred to as primary users), CR technology permits unlicenced users (usually referred to as CR users) to utilise idle bands as long as they do not cause harmful interference to primary users [2].

The operation of a CR network is more complicated than other wireless networks because the CR nodes dynamically access the available channels. Detecting the presence of primary users and further determining the availability of certain channels are regarded as a major technical challenge in CR networks [3]. Hence, spectrum sensing is considered as an important issue of CR networks that aim to find the vacant frequency bands in order to allow CR users access to licenced bands in an opportunistic manner [4].

According to the deployment scenario, CR networks can be classified into two basic types of networks: one is the infrastructure-based CR networks, and the second is the infrastructure-less CR networks [5]. In the infrastructure-based CR networks, all CR nodes directly communicate with the central network entity, which is responsible for managing the network operations, for instance, spectrum sensing and spectrum assignment [6]. On the other hand, in the infrastructure-less CR networks, also known as CR ad hoc network, no central entity is present. Therefore, CR nodes have to rely on themselves for spectrum sensing, assignment and management. The application of CR technology in distributed scenarios remains underdeveloped due to a lack empirical research [7].

Broadcasting is considered a fundamental operation in wireless and cognitive radio networks (CRNs). The operation of most network protocols in the ad hoc network depends on broadcasting control information among neighbouring nodes, such as spectrum sensing and routing information.

In traditional single-channel or multichannel ad hoc networks, due to uniform channel availability, broadcasting is easily implemented as nodes are tuned to a single common channel. On the contrary, broadcasting in CR ad hoc networks is a challenging task and much more complicated. The complexity emerges from the fact that different CR users might acquire different channels at different times. Consequently, this partitions the network into different clusters. Cognitive radio (CR) ad hoc networks rely on extensive exchange of control messages among neighbouring nodes to coordinate critical network functions such as cooperative sensing, routing, medium access, etc. To reliably broadcast these messages, a preassigned common control channel is needed. However, assigning a static control channel contradicts the opportunistic access nature of cognitive radio networks (CRNs).

In this chapter, the problem of broadcast in ad hoc CR networks is discussed, current solutions for the problem are reviewed and an intelligent solution for broadcasting based on graph theory to connect different local topologies is developed, which is a unique feature in CRNs. The remainder of this chapter is organised as follows: Section 2 describes the related work in this area. Then the broadcast problem is presented with the system model in Section 3. The proposed broadcast protocol for multi-hop CR ad hoc networks is presented in Section 4. Performance evaluation is conducted in Section 5, followed by conclusions in Section 6.

2. Related work for broadcasting in CR network

In the literature, several works have extensively studied the broadcasting issue in traditional ad hoc networks, Mobile Ad Hoc Networks (MANETs), Wireless Mesh Networks (WMNs), vehicular ad hoc networks (VANETs) and wireless sensor networks (WSNs). Nevertheless, there are a few studies that investigate the problem of broadcasting in CR ad hoc networks. These works propose numerous performance goals, for example, optimisation of throughput, delay and data delivery. However, most of these techniques cannot be used in practical scenarios due to their limitations and impractical assumptions.

In the recent literature, many protocols have been presented for exchanging messages in CR networks. One of the simplest suggestions is broadcasting over the unlicenced bands such as ultra-wide band (UWB) or industrial, scientific and medical (ISM) [8]. This proposal cannot guarantee the reliability because these unlicenced bands are already overcrowded. Since many wireless devices communicate in the same band, harmful interference may significantly degrade the performance of broadcasting.

The authors in Ref. [9] propose a new strategy for broadcasting. They classify the channels based on the primary radio (PR) vacancy and CR occupancy. This strategy transmits on a single channel; therefore, CR nodes within the transmission range of the sender may not be able to receive the transmitted data if they tune onto a different channel. In Ref. [10], the authors proposed that the secondary network composed of a set of single-antenna secondary receivers (SRs) and one multi-antenna secondary transmitter (ST). The main responsibility of ST is to broadcast the message to the SRs without interfering the primary communication. Since the secondary users use orthogonal beamforming techniques, they can access the licenced spectrum without causing an interference to primary transmission.

The use of a dedicated control channel has been proposed to enable control message exchanging in multichannel networks [11, 12]. To transmit or receive messages, the CR node must tune onto the common control channel (CCC). In CR networks, it is very difficult to find an idle common channel for all nodes. Hence, this technique is not considered to be feasible. Different schemes have been proposed for establishing a local common control channel for exchanging messages [13, 14]. However, most of these schemes require prior information about the set of idle channels across all the CR nodes in the network.

In Ref. [15], the authors assume that the same idle channels between CR nodes are a must to successfully broadcast data. The proposed approaches in Ref. [16] assume that the CR node hops across the channels based on a random channel-hopping sequence to transmit broadcast data. This scheme cannot guarantee reliable dissemination even if there is a common channel between nodes. In Ref. [17], the authors study the issue of broadcasting using multiple transceivers. It is assumed that the number of transceivers of each CR node is equal to the number of channels. This will raise the operational cost and the complexity of the CR device; therefore it is considered an impractical choice.

Many algorithms assume prior knowledge of the channel availability information and the global network topology [18, 19]. A time-efficient broadcast algorithm is presented in Ref. [18], where a set of channels and nodes are selected to convey a message from the source node to its neighbours. The authors in Ref. [19] propose a simple heuristic algorithm to transmit the messages between CR nodes in CR ad hoc networks. In this work, CR nodes are assumed to be equipped with multiple transceivers to broadcast to multiple channels.

3. Broadcast problem and system model

In this section, the broadcast problem, the system model and the basic assumptions are presented.

3.1. Broadcast problem in CR networks

To further illustrate the challenges associated with broadcast in CR ad hoc networks, consider simple single-hop broadcast topology for traditional and CR network shown in **Figures 1** and **2**, respectively, where *node A* is the source node with *N* neighbours.

In traditional ad hoc networks, all nodes can tune to the same channel due to the uniformity of channel availability. Therefore, broadcasting a message can be easily implemented over a single common channel as all nodes receive messages from the same channel. As shown in **Figure 1**, *node A* only needs to broadcast over a single channel to deliver the broadcast message to all its neighbouring nodes.

However, in CR ad hoc networks, the opportunity of a common channel available for all CR nodes may not exist. In addition, different CR users might acquire different channels at different times. Therefore, broadcasting in cognitive radio ad hoc networks is a much more challenging task. As shown in **Figure 2**, to deliver the broadcast message to all the neighbouring

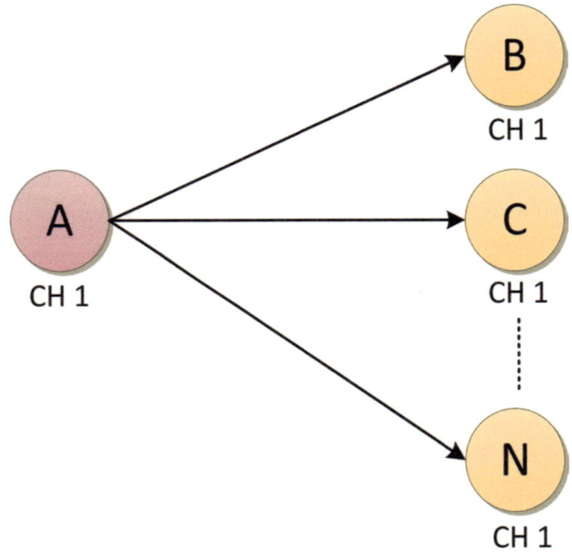

Figure 1. Single-hop broadcast topology for traditional network.

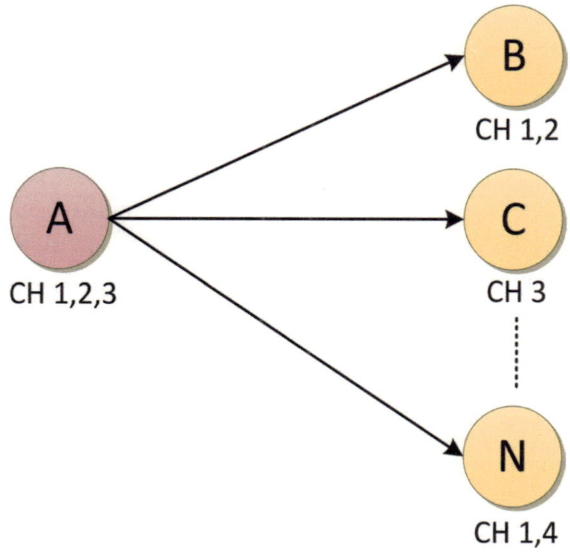

Figure 2. Single-hop broadcast topology for CR network.

nodes, node A needs to transmit the broadcast message to different channels. In the worst case, each neighbouring node may tune onto a different channel. Consequently, the source node has to broadcast over all the channels.

In fact, the reliable broadcasting in CR ad hoc networks depends on connecting different local topologies. Hence, the broadcast channel(s) should be carefully and dynamically allocated in order to secure a reliable communication in CR networks.

3.2. Network model

A CR ad hoc network with no centralised coordinator is considered. Hence, network environment tasks like channel selection, neighbour discovery and spectrum sensing are individually accomplished by the CR users.

We consider a set of N cognitive radio (CR) nodes $\{CR_1, CR_2, \ldots, CR_n\}$ and a set of M primary radio (PR) nodes $\{PR_1, PR_2, \ldots, PR_m\}$ in the same geographical area. Primary radio nodes are the licenced users, and they can access their respective licenced bands without any restriction. While CRs can access licenced bands opportunistically, i.e. they are allowed to use the idle licenced bands only if they do not interfere with ongoing PR transmissions.

Note that an idle state describes the temporal availability of a channel. To prevent interference, CR users are capable of sensing spectrum opportunities using energy detectors, cyclostationary feature extraction, pilot signals or cooperative sensing [5].

A set of K nonoverlapping orthogonal frequency channels ($C_{global} = C_1, C_2, \ldots, C_k$) is considered, which may be freely occupied by the PR users. Each CR node knows the global channel set C_{global} and can operate on a subset C_{local} of this global channel set depending on the local channel availability at that node, where $C_{local} \subseteq C_{global}$. For simplicity, it is assumed that all channels have the same capacity. However, the proposed protocol can be easily extended to channels of different capacities.

In this model, it is assumed that CR nodes are equipped with half-duplex transceivers that can either receive or transmit (not both) on a single channel at any given time. Each CR can swiftly hop between channels using software-defined radio (SDR) technology. The utilisation of a single transceiver reduces the operational cost of the CR device [6], as well as avoiding any potential interference between adjacent transceivers due to their close proximity [7].

Throughout this chapter, it is assumed that the channel availability is relatively stable (i.e. during a short period of time, channel status does not change). Therefore, the proposed protocol is more suited to the case of temporal underutilisation and spatial spectrum underutilisation when the activity of PR user is not very dynamic. The main notations used in the chapter are summarised in **Table 1** for easy reference.

3.3. Sensing spectrum holes

Spectrum sensing aims to identify the available spectrum and prevent any harmful interference to the primary users. It is assumed that CR nodes periodically perform spectrum sensing to ensure up-to-date information regarding the PR activity and identify the available channels.

In addition, it is assumed all CR nodes are synchronised and follow the same sensing cycles. In the sensing period, no transmission is allowed, and all CR nodes must be silent. Therefore, the time needed to deliver a packet in the network may be influenced when the CR nodes are banned from transmission due to the imposition of the silent duration.

The transmission time and the spectrum sensing for every CR user are T_t and T_s, respectively, where T_t is the effective duration of time for which transmission is allowed for any CR node

Symbols	Descriptions
N	Set of CR nodes
C_{global}	Total number of channels
C_i	The available channel set of CR_i
N_i	Set of single-hop neighbours of CR_i
T_s	Spectrum sensing time for CR users
T_t	Transmission time for CR users
$Cr^r{}_{PR}$	Transmission range of PR users
$Cr^r{}_{CR}$	Transmission range of CR users
$G(X, Y, E)$	Bipartite graph
BCS_i	Broadcast channel set of CR_i
TC_i	Tuning channel of CR_i

Table 1. Symbols used for OBA description.

on any choice of free spectrum, while T_s is the duration of time that all CR nodes must be silent for the purpose of sensing. $T_s + T_t$ gives the frame time for each user when considered together.

3.4. Discovering CR neighbouring nodes

To successfully deliver the broadcast messages to all the CR nodes in each neighbourhood, CRs must discover the network topology and the common idle channels that can be used to communicate among neighbours; these tasks are typically undertaken during the neighbour discovery.

In the absence of a common control channel, discovering neighbours in CR Ad Hoc Network (CRAHN) is undoubtedly a challenging task; we propose a neighbour discovery mechanism to address this issue. Initially, it is assumed that individual nodes are tuned to different channels and have no prior knowledge of their neighbours and the network topology. Furthermore, each CR node maintains the local idle channel list based on the information received from the spectrum sensing.

At the beginning of constructing the network, every CR node has to beacon its information (node's id and its available channels) onto all the locally available channels, one by one. As a result, all single-hop neighbours that are tuned to any idle channels are able to receive a copy of this message. Each CR node receives this beacon message and records the transmitter's CR node information in its single-hop neighbours list N_i. After forming and configuring the network, the CR nodes do not have to beacon messages unless there is a change in their channel availability.

4. Intelligent broadcasting algorithm

In this section, we present optimal broadcasting algorithm (OBA), the proposed broadcast protocol for ad hoc CR networks. OBA mainly aims to maximise the reachability and increase the reliability of data dissemination in ad hoc CR networks. To guarantee successful broadcast operation, the OBA protocol adapts to current network characteristics. Hence, based on local measurements of the PR activities, each CR node independently classifies the set of available channels. This process of classification is refined by determining the minimum set of broadcast channels. The receiving channel of a given node is identified from this minimum set of channels. This tuning channel is selected from the set because it has no PR activity, thus being able to reach a higher number of CR neighbours.

To increase the network connectivity, the OBA aims to converge CRs that possess similar spectrum opportunities to the same channel. This in turn reduces the delay in packet dissemination and number of transmissions over multiple channels. This aim is motivated by two key factors. First, grouping CRs with similar available channels indirectly initiates hard-decision cooperative sensing [8, 20]. Second, it minimises the required number of channels in the set that are needed to connect all neighbouring nodes.

In order to guarantee conveying the broadcast message to all the neighbouring nodes in each transmission, the CR sender broadcasts the message over a minimum set of the available channels that are shared between the sender and its neighbouring nodes.

CR nodes with no transmission requests refine the channels based on the same criterion to select the best channel for message reception. The use of similar refining techniques by all nodes in the network means it is highly probable that CR users in close geographic locations select the same channel set. It is probable that neighbouring CR nodes of the transmitter CR node sense the same primary activity. It is highly likely that channels at a CR node's disposition are also free to its neighbours [11]. Therefore, the probability of creating a connected topology is increased by OBA. Once a message is received, the intermediate CR node carries out the procedure again and rebroadcasts in order to deliver the message to its neighbours.

The following characteristics were considered crucial in the development of OBA: (i) decentralisation, distributed implementation of channel allocation; (ii) convergence, CR users with the same available channels individually converge to the same channel decisions; (iii) delay and communication efficiency, channel allocation is achieved with no exchange of messages; and (iv) adaptability, reallocation is required only in the case where there is a change in the network topology.

4.1. Primary radio activity model

The primary user activity on the licenced channels has a vital effect on the CR network performance. Therefore, the realisation of CR protocols depends on the estimation of these activities. In the literature, many works have used the PR activity model [5, 9, 21], where the PR traffic is modelled as an alternating renewal process of *idle* (*OFF*) and busy (*ON*) periods. It is assumed that both *OFF* and *ON* periods of the PR activity are independent and identically distributed (*i.i.d.*). In addition, the alternating renewal process is represented by a two-state birth-death process with birth rate λ_{on} and death rate λ_{off} [22].

Let $1/\lambda_{on}$ and $1/\lambda_{off}$ be the average *ON* and *OFF* times of the *k*th channel. The probability of the *k*th channel being occupied is given by

$$P^k_{busy} = \frac{\lambda_{off}}{\lambda_{on} + \lambda_{off}}, \tag{1}$$

where $1 \leq k \leq K$ (the total number of channels). Therefore, the probability of utilising the *k*th channel (i.e. the channel being idle) without causing harmful interference to the primary users is

$$P^k_{idle} = 1 - P^k_{busy} = \frac{\lambda_{on}}{\lambda_{off} + \lambda_{on}}. \tag{2}$$

Let the channel set that match the user requirements (i.e. probability of idle channels being greater or equal to a predefined threshold P_{th}), represented by Φ. From Eq. (2), the set of available channels Φ for each node can be obtained as follows:

$$\Phi^k_{idle} = P_{th}, \quad \forall k \in \Phi, \; 1 \leq k \leq K \tag{3}$$

4.2. Bipartite graph representation

Different CR nodes may detect different available channel sets and neighbours due to the temporal and spatial distribution of the primary activity. To initiate communication with other nodes, each CR node must construct a local view of the network topology. This will include neighbouring nodes in the vicinity and their channels' information. A bipartite graph jointly models the set of shared idle channels among neighbouring nodes and the CR network topology [23].

A graph $G(V, E)$ is called a bipartite graph $G(X, Y, E)$ if the set of vertices V can be divided into two disjoint sets X and Y with $X \cup Y = V$, such that each edge in E has one endpoint in X and the other in Y.

Each CR_i detects available channel set λ^k_i and acquires information from its single-hop neighbours N_i on their available channels. To jointly represent the similarities between its own idle channels and those of its neighbours, it can construct an undirected bipartite graph. Each CR_i constructs a bipartite graph $G_i(X_i, Y_i, E_i)$, where the single-hop neighbour N_i is represented by the set of vertices X_i, and while the set of channels λ^k_i is represented by the set of vertices Y_i.

An edge (x, y) exists between vertex $x \in X_i$ and a vertex $y \in Y_i$ if and only if $y \in \lambda^k_i$, i.e. channel y is in the idle channel set of both CR_i and CR_j. Note that CR_i is connected to all vertices in Y_i, since $Y_i = \lambda^k_i$. The graph model is then used as the basis for computing the broadcast channel set.

Figure 3 shows the topology graph for a CRN with six nodes. **Figure 4** shows the bipartite graph $G_A(X_A, Y_A, E_A)$ constructed by CR_A. **Figure 5** presents the bipartite graph $G_B(X_B, Y_B, E_B)$ constructed by CR_B, for the same topology of **Figure 3**. Note that $G_A \neq G_B$ despite the fact that CR_A and CR_B are one-hop neighbours. This holds true because $N_A \neq N_B$ and with different physical locations, it is expected that $\lambda^k_A \neq \lambda^k_B$.

4.3. Broadcast channel selection

Based on its own bipartite graph $G_i(X_i, Y_i, E_i)$, the CR determines the minimum broadcast channel set (BCS) and selects the finest channel as the tuning channel (TC). The problem of determining the minimum broadcasting channels set for a CR node can be modelled as the set cover problem.

The set cover problem is defined as follows: Given a set of n elements called the universe $U = \{U_1, U_2, \ldots, U_n\}$ and a set of m subsets of U, $S = \{S_1, S_2, \ldots, S_m\}$, find a minimum collection C of sets from S such that C covers all elements in U [24].

Finding the minimum and most effective BCS is the main goal of the OBA. Therefore, it represents the universe by the set of vertices X, the sets by the set of vertices Y and the inclusion of elements in sets as edges. Thus, Y has been transformed into a set of subsets of X. The aim is to identify the minimum cardinality subset of Y that covers all vertices of X. Finding the minimum set cover is an NP-complete problem [25].

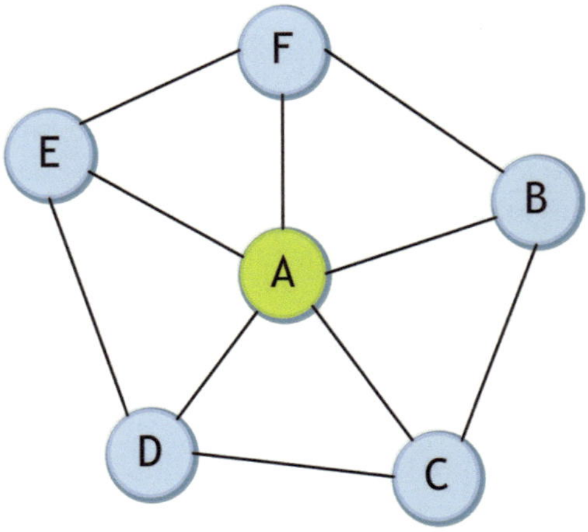

Figure 3. Six-node CR network.

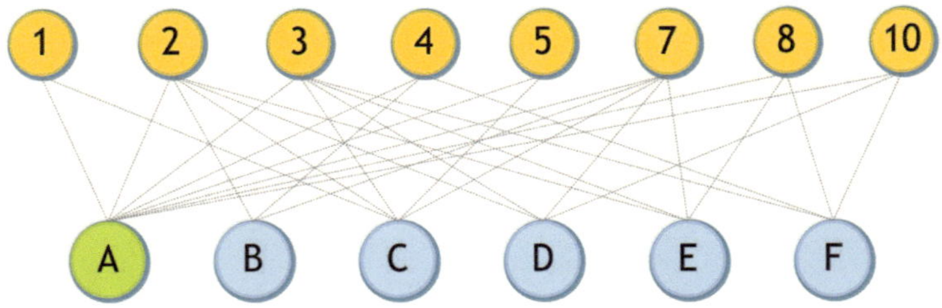

Figure 4. Bipartite graph constructed by node A.

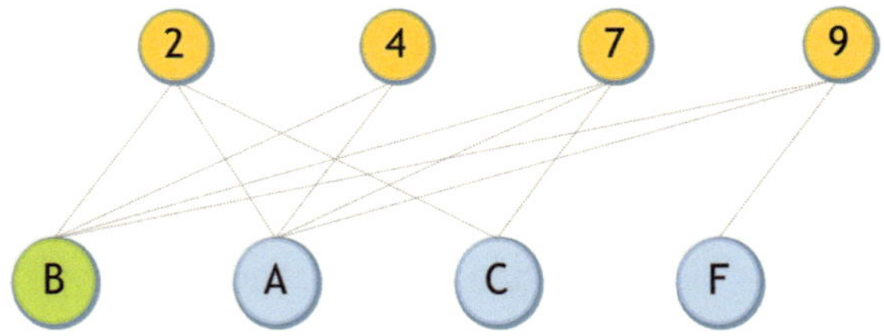

Figure 5. Bipartite graph constructed by node B.

A comprehensive search may be possible for bipartite graphs of small size. However, the space of possible solutions grows exponentially with the cardinality of the vertex set. Hence, OBA has been introduced as it is a greedy heuristic algorithm for finding a cover set with a minimum number of channels.

OBA continually examines a single channel. In each round, OBA selects the channel which connects the greatest number of nodes that have not been covered yet. The indexes of channels that have already been chosen is represented by vector BCS, while the set of CRs that are not covered yet by the channels in the BCS is represented by U. Initially, BCS = 0, while U = Xi and S = Yi. In each round, a channel Si from S is chosen, which has the maximum overlap with U.

Then, Si will be added to BCS, removed from S, and any CR users covered by Si will be dropped from U, and the operation will be repeated until U is empty.

Broadcast channels set BCS as the output. It is sorted in descending order based on the number of neighbours covered per channel. Prioritising channels in descending order is essential for two key reasons: (1) to guarantee the node picks the first channel in the list as the tuning channel TC enabling the maximum connectivity with its neighbouring nodes and (2) the node will ensure that the maximum number of neighbouring nodes are targeted during the first transmission, the second highest number of neighbours will be targeted during the second transmission and so on. **Figure 6** explains the operation of the OBA algorithm.

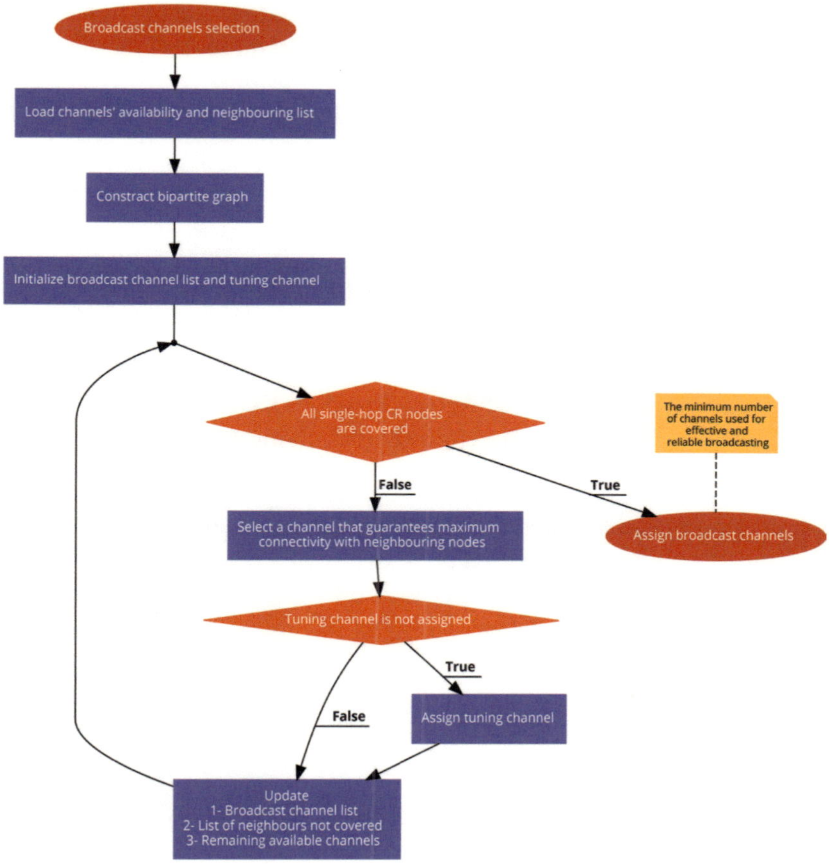

Figure 6. Flowchart for OBA.

5. Performance evaluation

The OBA protocol is implemented with the ns-2 simulator; we randomly deployed 100 CR nodes in a square region of sides 1000 m. The sensing and transmission times are set to $T_s = 0.1$ s and $T_t = 0.6$ s, respectively. To get accurate results, we repeated each group of simulations

100 times. The transmission range of each CR user is set to CrCR = 150 m. Moreover, the PR user has a transmission range CrPR = 250 m.

The performance of the proposed OBA protocol is compared against three studies: (a) SURF strategy proposed in Ref. [9], (b) selective broadcasting (SB) presented in Ref. [19] and (c) random strategy (RS).

In SURF, the channels are classified based on the PR vacancy and CR occupancy. This scheme transmits on a single channel; therefore, CR nodes within the transmission range of the sender may not be able to receive the transmitted data if they tune onto a different channel.

Instead of broadcasting over a single channel, in SB CR nodes broadcast the information over a selected group of channels. This approach requires the node to tune to more than one channel. SB does not provide any synchronisation between transmitter and receiver nodes.

However, RS strategy has been chosen as it is the simplest, and no further information is required. In RS, CR nodes randomly pick channels for transmission and/or receiving, without any consideration of the ongoing CR and PR activity over these channels.

5.1. PR communication protection

In this section, probable interference ratio (PIR) is characterised. Due to an inappropriate channel decision from OBA, RS, SB and SURF, PIR is caused by CR nodes to PR nodes. **Figure 7** demonstrates that OBA allows less interference to PR nodes, compared to SURF, SB and RS. The proposed broadcasting protocol (OBA) tries to reduce the interference with the PR users' communication. This is achieved by intelligently identifying the unutilised spectrum based on a real-time sensing.

In OBA, if there is no available channel at the time of broadcasting due to PR activity, the CR will not broadcast the message. **Figure 4** illustrates a small PIR value for OBA, which demonstrates the cases where potential interference would happen if no channel was available and the broadcasting continued.

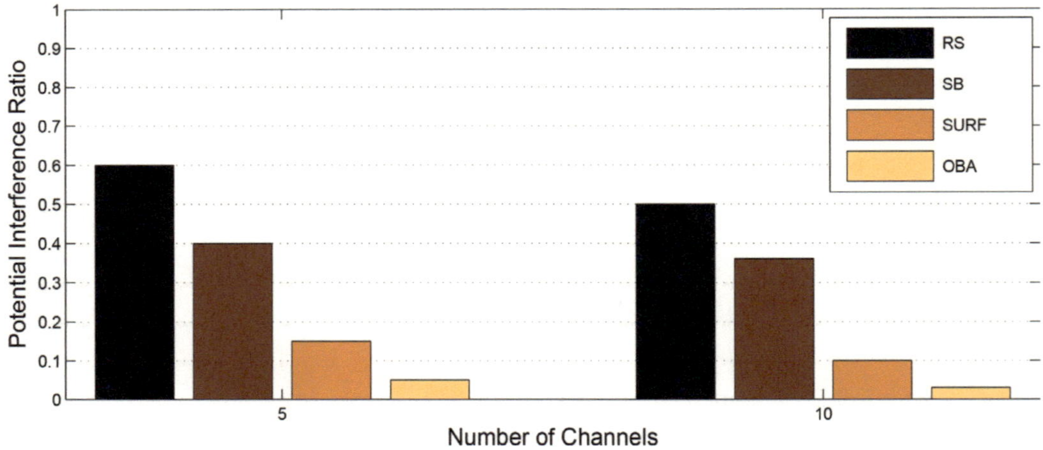

Figure 7. The effect on the PR users due to CR transmissions.

5.2. Packet delivery ratio

Table 2 and **Figures 8** and **9** show the Packet Delivery Ratio (PDR) of OBA, SURF, SB and RS, when the total number of channels (Ch) is Ch = 5 and Ch = 10, respectively. Compared to other schemes, OBA performed very well and achieves a significant packet delivery ratio.

OBA ensures approximately a 70–80% successful delivery ratio when Ch = 5, while in the case of SURF, it is 32%, 21% in SB and 3% in RS. When Ch = 10, OBA guarantees almost an 80–90% successful delivery ratio, compared to 29% for SURF, 17% for SB and almost 1% in the case of RS.

It is important to mention that the diversity of PR activity and channel availability lead to the formation of different clusters (network topologies) at each CR node. To overcome this problem, OBA creates communication links with other clusters by broadcasting over the minimum set of idle channels. This covers all CR users in the vicinity that increase the successful delivery of the broadcast messages.

RS does not ensure the broadcast channel is free from PR activity for its transmission. Therefore, this causes a severe reduction in the delivery ratio. Although broadcasting is performed using multiple channels, SB achieves less successful broadcast delivery compared to OBA.

In certain cases, because of the lack of transmitter/receiver synchronisation between nodes (i.e. the tuning channel is selected randomly), the transmitter may not have any effective receivers. SURF strategy transmits using a single channel. This means only CR nodes within the transmission range of the source node and tuned to the same channel will be able to receive the broadcast message.

It is worth noting that the performance of OBA is slightly enhanced when the number of channels is increased. Since adding more channels automatically spreads nodes over more channels; thus, this result is not unreasonable. However, OBA achieves better results when more channels are available, when the proper metric is used and the same algorithm is employed at the sender and the receiver.

5.3. Channel set size

The number of channels utilised by a CR node to broadcast a message to its single-hop neighbouring node is defined as the channel set size (CSS). In **Figure 10**, the CSS of OBA is compared with the CSS of RS, SB and SURF in relation to the number of available channels. It is clear that SB utilises nearly all of the available channels.

Broadcast technique	Packet delivery ratio when Ch = 5 (%)	Packet delivery ratio when Ch = 10 (%)
Optimal broadcasting algorithm (OBA)	76	83
SURF	32	29
Selective broadcasting (SB)	21	17
Random strategy (RS) 32%	3	1.2

Table 2. Successful packet delivery ratio.

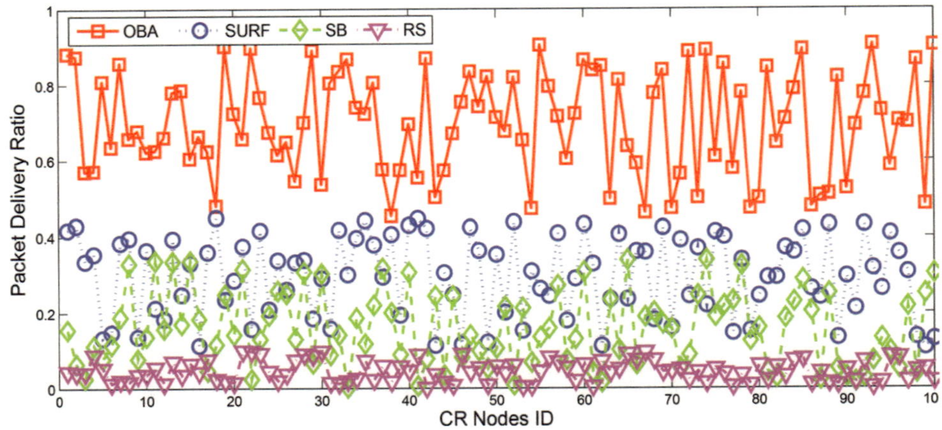

Figure 8. Successful packet delivery ratio, Ch = 5.

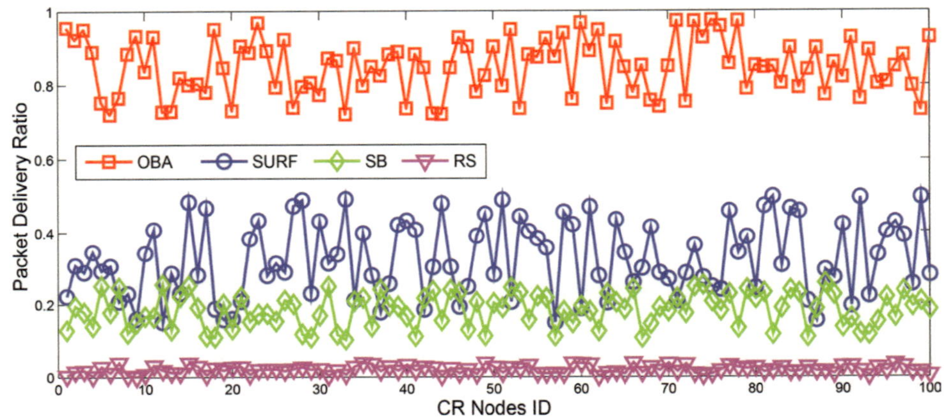

Figure 9. Successful packet delivery ratio, Ch = 10.

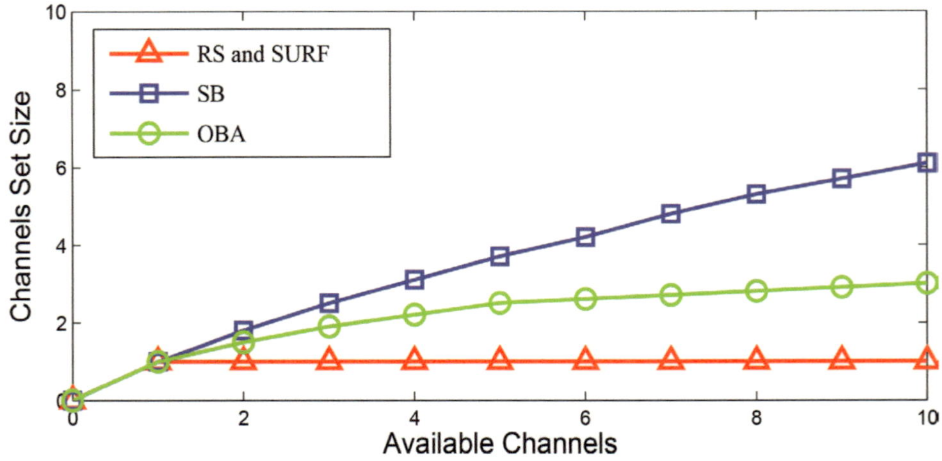

Figure 10. Average number of used channels for broadcast per node.

Moreover, the number of the utilised channels increases when there is more available channels. This can be explained by the fact that CR nodes are spread over more channels when the number of idle channels increases. Consequently, the CR user needs to broadcast over more channels. However, the CSS in the case of OBA is considerable.

Furthermore, the increase in the number of available channels does not significantly impact the number of used channels. This is achieved by OBA and by using proper metrics to prevent CR nodes from dispersing over all the available channels. In addition, OBA helps to merge the neighbouring CR nodes to the same channel selection. This in turn results in a considerable reduction in the CSS.

Irrespective of the number of available channels, RS and SURF use only a single channel for the transmission. It is difficult to use a single channel for broadcasting in CR networks. This is because of the nonuniform channel availability and the impossibility of a global common channel being available. In the case of OBA, most of the CR neighbouring nodes will successfully receive the broadcasting message. This is because OBA connects different local topologies, which results in a significant increase in the successful packet delivery.

6. Conclusion

In this chapter, the problem of reliable broadcast in ad hoc CRNs is addressed. Due to the spatial variation in spectrum availability, different CR nodes might sense different idle channels, which can partition the network into different clusters. By jointly representing the spectrum availability and the network local topology as a bipartite graph, the problem of connecting different CR nodes can be mapped to the problem of set cover. An intelligent algorithm is developed, which guarantees a distributed reliable selection of the broadcast channel with a facilitative channel switching facility where primary user activity is detected without the need for frequent reselection. It has been shown through simulation results that the only way to provide a reliable broadcast that considers the spatial variation of spectrum availability is through connecting different local topologies in the absence of a common control channel.

Acknowledgements

Part of this chapter has been taken from our earlier work cited in Ref. [21].

Author details

Yasir Ahmed Al-Mathehaji*, Said Boussakta and Martin Johnston

*Address all correspondence to: y.a.a.al-mathehaji@ncl.ac.uk

School of Electrical and Electronic Engineering, Newcastle University, Newcastle upon Tyne, UK

References

[1] Tadayon N, Aissa S. Modeling and analysis framework for multi-interface multi-channel cognitive radio networks. IEEE Transactions on Wireless Communications. 2015;**14**:935-947

[2] Al-Mathehaji Y, Boussakta S, Johnston M, Hussein J. Primary receiver-aware opportunistic broadcasting in cognitive radio ad hoc networks. In: Eighth International Conference on Ubiquitous and Future Networks (IEEE-ICUFN); July 2016. pp. 30-35

[3] Li H, Cheng X, Li K, Hu C, Zhang N, Xue W. Robust collaborative spectrum sensing schemes for cognitive radio networks. IEEE Transactions on Parallel Distributed Systems. Aug 2014;**25**:2190-2200

[4] Luis M, Oliveira R, Dinis R, Bernardo L. Characterization of the opportunistic service time in cognitive radio networks. IEEE Transactions on Cognitive Communications and Networking. Sept 2016;**2**:288-300

[5] Chowdhury KR, Akyildiz IF. Crp: A routing protocol for cognitive radio ad hoc networks. IEEE Journal on Selected Areas in Communications. 2011;**29**:794-804

[6] Tukmanov A, Boussakta S, Ding Z, Jamalipour A. Outage performance analysis of imperfect-csi-based selection cooperation in random networks. IEEE Transactions on Communications. Aug 2014;**62**:2747-2757

[7] Akyildiz IF, Lee W-Y, Chowdhury KR. Crahns: Cognitive radio ad hoc networks. Ad Hoc Networks Journal. 2009;**7**:810-836

[8] Brown TX. An analysis of unlicensed device operation in licensed broadcast service bands. In: First IEEE International Symposium on Dynamic Spectrum Access Networks. IEEE; 2005. pp. 11-29

[9] Rehmani MH, Viana AC, Khalife H, Fdida S. Surf: A distributed channel selection strategy for data dissemination in multihop cognitive radio networks. Computer Communications. 2013;**36**:1172-1185

[10] Chraiti M, Hakim H, Ajib W, Boujemaa H. Spectrum sharing techniques for broadcast cognitive radio networks. IEEE Transactions on Wireless Communications. 2013;**12**(11): 5880-5888

[11] Campolo C, Molinaro A, Vinel A, and Zhang Y. Modeling prioritized broadcasting in multichannel vehicular networks. IEEE Transactions on Vehicular Technology. 2012;**61**(2): 687-701

[12] Chen J and Chen Y-D. Amnp: Ad hoc multichannel negotiation protocol for multihop mobile wireless networks. In: IEEE International Conference on Communications; 2004. IEEE; 2004. pp. 3607-3612

[13] Ghasemi A, Sousa ES. Opportunistic spectrum access in fading channels through collaborative sensing. Journal of Communications. 2007;**2**(2):71-82

[14] Bian K, Park J-M, Chen R. Control channel establishment in cognitive radio networks using channel hopping. IEEE Journal on Selected Areas in Communications. April 2011;**29**:689-703

[15] Song Y, Xie J. A distributed broadcast protocol in multi-hop cognitive radio ad hoc networks without a common control channel. In: IEEE INFOCOM; March 2012. pp. 2273-2281

[16] Theis N, Thomas R, DaSilva L. Rendezvous for cognitive radios. IEEE Transactions on Mobile Computing. Feb 2011;**10**:216-227

[17] Qadir J, Misra A, Chou CT. Minimum latency broadcasting in multi-radio multi-channel multi-rate wireless meshes. In: IEEE International Conference on Sensing, Communication and Networking; 2006. pp. 80-89

[18] Arachchige CJL, Venkatesan S, Chandrasekaran R, Mittal N. Minimal time broadcasting in cognitive radio networks. Springer Journal on Distributed Computing and Networking. 2011:364-375

[19] Kondareddy YR and Agrawal P. Selective broadcasting in multi-hop cognitive radio networks. In: IEEE Sarnoff Symposium; 2008. pp. 1-5

[20] Resta G, Santi P, Simon J. Analysis of multi-hop emergency message propagation in vehicular ad hoc networks. In: The 8th ACM International Symposium on Mobile Ad Hoc Networking and Computing. ACM; 2007. pp. 140-149

[21] Al-Mathehaji Y, Boussakta S, Johnston M, Fakhrey H. Crbp: A broadcast protocol for cognitive radio ad hoc networks. In: IEEE International Conference on Communications (ICC); June 2015. pp. 7540-7545

[22] Yuan G, Grammenos R, Yang Y, and Wang W. Performance analysis of selective opportunistic spectrum access with traffic prediction. IEEE Transactions on Vehicular Technology. May, 2010;**59**:1949-1959

[23] Lazos L, Liu S, Krunz M. Spectrum opportunity-based control channel assignment in cognitive radio networks. In: 6th Annual IEEE International Conference on Sensing, Communication and Networking (SECON); 2009. pp. 1-9

[24] Hochbaum DS. Approximation algorithms for the set covering and vertex cover problems. SIAM Journal on Computing. 1982;**11**(3):555-556

[25] Karp RM. Reducibility among combinatorial problems. In: Complexity of Computer Computations. Springer US; 1972. pp. 85-103

Chapter 3

Outage Performance Analysis of Underlay Cognitive Radio Networks with Decode-and-Forward Relaying

Mustafa Namdar and Arif Basgumus

Additional information is available at the end of the chapter

http://dx.doi.org/10.5772/intechopen.69244

Abstract

In this chapter, we evaluate the outage performance of decode-and-forward relaying in cognitive radio networks over Rayleigh fading channels, subject to the relay location for a secondary user. In particular, we obtain the optimal relay location in wireless communications systems for the cognitive radio networks, using differential evolution optimization algorithm. Then, we investigate the optimal transmission rate of the secondary user. We present the numerical results to validate the proposed theoretical analysis and to show the effects of the Rayleigh fading channel parameters for the whole system performance.

Keywords: cognitive radio networks, decode-and-forward relaying, differential evolution optimization algorithm, optimal relay location, outage probability

1. Introduction

Cognitive radio (CR) is a new approach for wireless communication systems to utilize the existing spectrum resources efficiently. Spectrum utilization can be increased by opportunistically allowing the unlicensed secondary user (SU) to utilize a licensed band in the absence of the primary user (PU) [1–4]. The ability of providing awareness about the usage of the frequency spectrum or the detection of the PU in a desired frequency band lets the SU access the radio communication channel without causing harmful interference to the PU [5–8].

Cooperative wireless communications, which depend on cooperation among distributed single-antenna wireless nodes, have emerged recently as an alternative to multi-antenna systems to obtain spatial diversity [9–13]. In a wireless communication system, when the source terminal does not have a good-enough link with the destination one, cooperative relaying can be utilized to improve spectral efficiency, combat with the effects of the channel fading and to increase the channel capacity. There are various cooperative relaying schemes and two of the most widely studied in the literature are amplify-and-forward (AF) and decode-and-forward (DF) protocols. Between them, the DF cooperation protocol is considered in this chapter, in which the relay terminal decodes its received signal and then re-encodes it before transmission to the destination [14]. In order to achieve higher outage performance, we investigate the DF relaying in CR networks over Rayleigh fading channels, subject to the relay location for a SU. Then, we obtain the optimal relay location for the CR networks and optimal transmission rate of the SU using the differential evolution (DE) optimization algorithm [15–17].

Most of the previous publications have studied the performance of cooperative communications techniques over different fading channels and under different constraints [18–26]. In [18], the authors derive the analytical error rate expressions to develop power allocation, relay selection and placements with generic noise and interference in a cooperative diversity system employing AF relaying under Rayleigh fading. Woong and Liuqing [19] address the resource allocation problem in a differentially modulated relay network scenario. It is shown to achieve the optimal energy distribution and to find optimal relay location while minimizing the average symbol error rate. The effect of the relay position on the end-to-end bit error rate (BER) performance is studied in [20]. Furthermore, Refs. [21–26] investigate the relay node placements minimizing the outage probability where the performance improvement is quantified. Although cooperative transmissions have greatly been considered in the above manuscripts, to the best of the our knowledge, there has not been any notable research for the relay-assisted CR networks based on the DE optimization algorithm. As far as we know, DE optimization algorithm has not been applied for obtaining the optimal location of the relaying terminal in CR networks over Rayleigh fading channels.

In summary, to fill the above-mentioned research gap, we here provide an optimization analysis yielding the optimal location of the relaying terminal for the SU in CR networks. Furthermore, we analyse the transmission rate for the SU over Rayleigh fading channels using DE optimization algorithm. As far as we know, DE optimization algorithm has not been applied for obtaining the optimal location of the relaying terminal and the transmission rate in CR networks over Rayleigh fading channels.

The rest of the chapter is organized as follows: the system model and performance analysis are described in Section 2 presenting the relay-assisted underlay cognitive radio networks. The numerical results and simulations are discussed in Section 3 with the DE optimization approach. Finally, Section 4 provides the concluding remarks.

2. System model and performance analysis

This section presents the system model for the CR networks with DF cooperative relaying protocol shown in **Figure 1**. We consider the method developed in [27] that the transmission links between the source-to-relay and relay-to-destination are subject to Rayleigh fading. In the

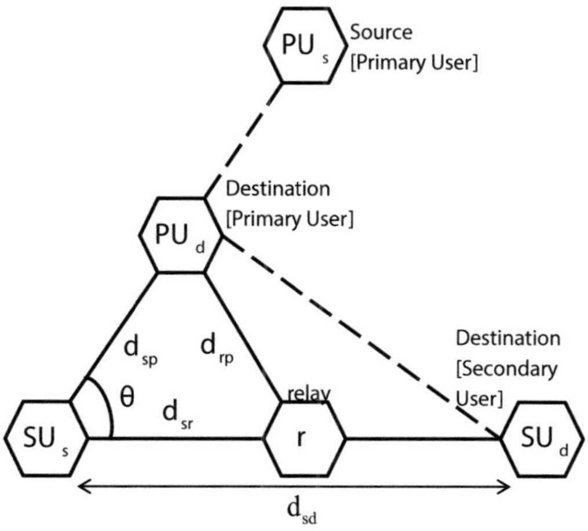

Figure 1. System model for cooperative relaying in cognitive radio networks [27].

system model for the cooperative relaying, we have both PU and SU, each with a source (PU_s and SU_s) and destination (PU_d and SU_d) nodes. Besides, the relay (r) is located in the same line between SU_s and SU_d. We assume that PU_s only transmits to the PU_d and SU_s utilize a two-phase cooperative transmission protocol causing interference to PU within a tolerable level. We also assume that equal-time allocation is implemented in the relayed transmission. In the first phase, SU_s transmits the signal to r. In the second phase of this transmission, r decodes its received signal and retransmits (forwards) it to the SU_d [27]. We denote the distance between the secondary source SU_s and the relay r as d_{sr}, the distance between the secondary source SU_d and the primary destination PU_d as d_{sp}, the distance between the secondary source SU_s and the secondary destination SU_d as d_{sd} and finally, the distance between the relay r and the primary destination PU_d as d_{rp}. We have

$$d_{rp}^2 = d_{sp}^2 + d_{sr}^2 - 2d_{sp}d_{sr}\cos\theta \tag{1}$$

where the cosine theorem is used. Here, θ is the angle between the horizontal axis and the line connecting the PU_d and SU_s nodes.

In a cognitive radio network, the transmission of a primary user has to be protected from the interference caused by either a secondary user or a relay. The level of the interference induced on the primary user (P_0) must be kept below a maximum tolerable level. On the other hand, when the level of interference from the secondary user's activity in the first phase or the relay transmission in the second phase exceeds the prescribed limit of P_0, this situation results in a corruption in the transmission of the primary user. Thus, the transmitting power levels of the primary user and relay have to be controlled and must not exceed P_0. Also, the outage probability of the primary destination during the source and relay transmission phases must be equal to a certain predetermined value such as ε_P. As the maximum transmitting power levels depends on the location of the relay, SU_s and ε_P, on the other hand, to maximize the data rate at the destination subject to the outage probability constraints, ε_s is evaluated by the secondary user.

Here, we consider the worst case channel conditions, namely, Rayleigh fading, might cause some signal power loss between the $SU_s - r$ and $r - SU_d$ links, also assuming N_0, power spectral density for the background noise is similar in the whole environment for the presented system model. In the literature, the outage probabilities for the PU_d during the source and the relay transmission phase are respectively given by $P_{out,source} = \exp\left(-P_o/P_s d_{sp}^{-\alpha}\right)$ and $P_{out,relay} = \exp\left(-P_o/P_r d_{rp}^{-\alpha}\right)$ where P_s is the transmit power of the SU_s and P_r is the transmit power of the relay, r [27]. It is assumed that these equations are equal to one another in order to maximize the transmission rate, and thus, the transmit powers for the secondary user and the relay are given as

$$P_s = \frac{P_0 d_{sp}^{\alpha}}{-\ln(\varepsilon_p)} \quad (2)$$

$$P_r = \frac{P_0 d_{rp}^{\alpha}}{-\ln(\varepsilon_p)} \quad (3)$$

respectively [27]. Here, α is the path loss exponent, and $\ln(.)$ is the natural logarithm operator.

In this study, it is aimed to minimize the outage probability of the secondary user for the DF relaying scheme and to maximize the transmission rate, R subject to the outage constraints of the primary user. The main objective of the proposed optimization algorithm is to find the optimal relay location on the direct link between SU_s and SU_d terminals. The outage probability of the secondary user for the DF relaying can be expressed as follows [27]:

$$\begin{aligned} P_{out} &= \left(1 - \exp\left(-\frac{g(R)}{2\overline{\gamma}_{sd}}\right)\right)\left(1 - \exp\left(-\frac{g(R)}{\overline{\gamma}_{sr}}\right)\right) \\ &+ \left(1 - \left(\frac{\overline{\gamma}_{sd}}{\overline{\gamma}_{sd} - \overline{\gamma}_{rd}}\exp\left(-\frac{g(R)}{\overline{\gamma}_{sd}}\right) + \frac{\overline{\gamma}_{rd}}{\overline{\gamma}_{rd} - \overline{\gamma}_{sd}}\exp\left(-\frac{g(R)}{\overline{\gamma}_{rd}}\right)\right)\right)\exp\left(-\frac{g(R)}{\overline{\gamma}_{sr}}\right) \end{aligned} \quad (4)$$

where R is the transmission rate for SU_s and $g(R) = 2^{2R} - 1$. We have

$$R = \frac{1}{2}\log_2\left(1 + \mu\sqrt{\varepsilon_s}\sqrt{\left(\left(\frac{d_{sd}}{d_{sp}}\right)^{-\alpha}\left(\frac{d_{rd}}{d_{rp}}\right)^{-\alpha}\left(\frac{d_{sr}}{d_{sp}}\right)^{-\alpha}\right)/\left(\left(\frac{d_{rd}}{d_{rp}}\right)^{-\alpha} + \left(\frac{d_{sr}}{d_{sp}}\right)^{-\alpha}\right)}\right). \quad (5)$$

Here, the outage probability for the secondary user is given by $\varepsilon_s = \left(\frac{1}{\overline{\gamma}_{sr}} + \frac{1}{\overline{\gamma}_{rd}}\right)\frac{1}{2\overline{\gamma}_{sd}} g(R)^2$. The average signal-to-noise ratios in the links PU_s to PU_d, SU_s to r, and r to SU_d are given by $\overline{\gamma}_{sd} = \mu\left(d_{sd}/d_{sp}\right)^{-\alpha}$, $\overline{\gamma}_{sr} = \mu\left(d_{sr}/d_{sp}\right)^{-\alpha}$, and $\overline{\gamma}_{rd} = \mu\left(d_{rd}/d_{rp}\right)^{-\alpha}$. We have $\mu = P_0/\left(-N_0 \ln(\varepsilon_p)\right)$.

For the optimization problem, a function is employed to minimize the outage probability and maximize the transmission rate for the DF relay-assisted CR system. DE optimization algorithm results show that the system performance can be significantly improved for the optimal value of the system parameters, seen in the following section.

http://dx.doi.org/10.5772/intechopen.69244

3. Numerical results and simulations

In this section, the numerical results are illustrated through the performance analysis curves of the proposed underlay cognitive radio networks with DF relaying. The detailed optimization results with the DE algorithm for DF relaying scheme are listed in **Table 1**. Here, the results for the optimal transmission distances, between secondary user source to relay ($SU_s - r$), $d_{sr_{opt}}$ are provided with different θ values, while $d_{sp} = d_{sd}$, $d_{sp} = 2\, d_{sd}$ and $d_{sp} = 5\, d_{sd}$. Besides, the maximum transmission rate values (R_{max}) for the secondary user, SU_s, are also illustrated in the same table. The results demonstrate that maximum transmission rate performance of the considered system increases while θ and d_{sp} increases.

The outage probability (P_{out}) performance of the considered system is illustrated in **Figure 2** with varying θ values when $(P_o/N_0) = 10$ dB, $\alpha = 4$, $\varepsilon_S = 0.1$, $\varepsilon_P = 0.05$, $d_{sp} = 2\, d_{sd}$ and $d_{sr} = d_{sd}/2$. It can be observed from the simulation results in **Figure 2** that the optimal θ angle can be calculated, where the best minimum of P_{out} is achieved.

$d_{sp} = d_{sd}$			$d_{sp} = 2d_{sd}$			$d_{sp} = 5d_{sd}$		
θ (°)	$d_{sr_{opt}}$	R_{max}	θ (°)	$d_{sr_{opt}}$	R_{max}	θ (°)	$d_{sr_{opt}}$	R_{max}
10	0.8830	0.5825	10	0.5295	2.7317	10	0.5042	5.4225
20	0.7606	0.6666	20	0.5276	2.7367	20	0.5039	5.4232
30	0.6819	0.7432	30	0.5246	2.7447	30	0.5037	5.4243
40	0.6261	0.8110	40	0.5206	2.7552	40	0.5030	5.4258
50	0.5835	0.8715	50	0.5160	2.7677	50	0.5024	5.4276
60	0.5497	0.9254	60	0.5109	2.7814	60	0.5017	5.4297
70	0.5222	0.9737	70	0.5055	2.7959	70	0.5009	5.4319
80	0.4995	1.0166	80	0.5001	2.8106	80	0.5000	5.4344
90	0.4807	1.0547	90	0.4949	2.8250	90	0.4992	5.4368
100	0.4651	1.0882	100	0.4899	2.8387	100	0.4983	5.4393
110	0.4521	1.1173	110	0.4853	2.8514	110	0.4975	5.4417
120	0.4414	1.1422	120	0.4812	2.8629	120	0,4967	5.4439
130	0.4328	1.1631	130	0.4777	2.8729	130	0.4960	5.4458
140	0.4259	1.1800	140	0.4747	2.8813	140	0.4954	5.4475
150	0.4207	1.1931	150	0.4724	2.8880	150	0.4950	5.4489
160	0.4171	1.2024	160	0.4707	2.8928	160	0.4946	5.4499
170	0.4149	1.2080	170	0.4697	2.8957	170	0.4944	5.4505
180	0.4142	1.2098	180	0.4694	2.8966	180	0.4943	5.4507

Table 1. Optimization results for DF relaying with different θ values for $d_{sp} = d_{sd}$, $d_{sp} = 2\, d_{sd}$, and $d_{sp} = 5\, d_{sd}$.

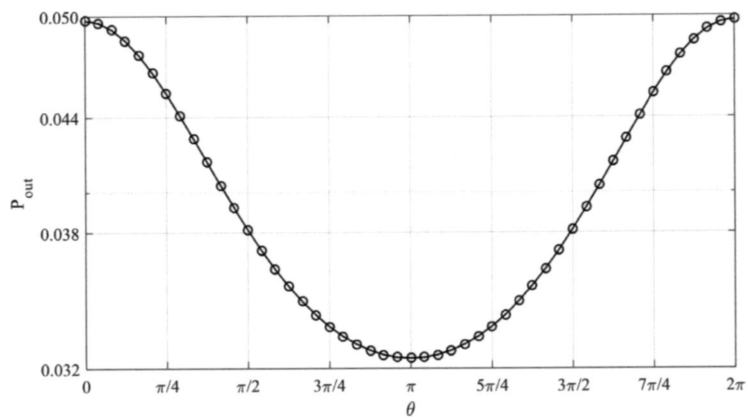

Figure 2. P_{out} for the considered underlay CR network with DF relaying under different θ values.

Figure 3 shows the transmission rate over Rayleigh fading channel versus (P_o/N_0) when $\alpha = 4$, $\varepsilon_S = 0.1$, $\varepsilon_p = 0.05$, $\theta = \pi/2$, $d_{sp} = 2\,d_{sd}$ and $d_{sr} = d_{sd}/2$. The results clearly show that R increases with the increase of the (P_o/N_0).

The transmission rate (R) of the considered system for the $SU_s - r$ link with the normalized d_{sd} distance is illustrated in **Figure 4** when $(P_o/N_0) = 10$ dB, $\alpha = 4$, $\varepsilon_S = 0.1$, $\varepsilon_p = 0.05$, $\theta = \pi/2$ and $d_{sp} = 2\,d_{sd}$. **Figure 4** indicates that the maximum transmission rate is achieved when the optimal transmission distances are used.

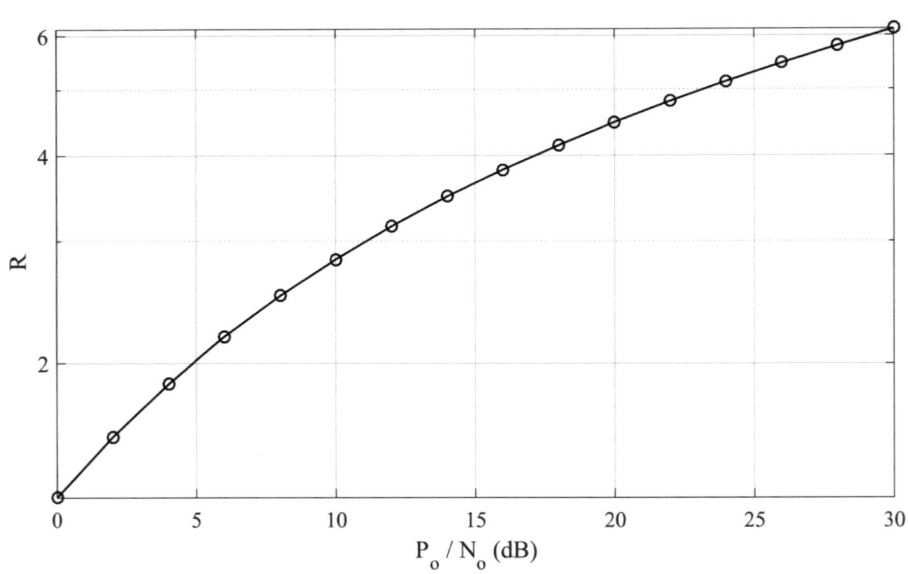

Figure 3. R vs. (P_o/N_0).

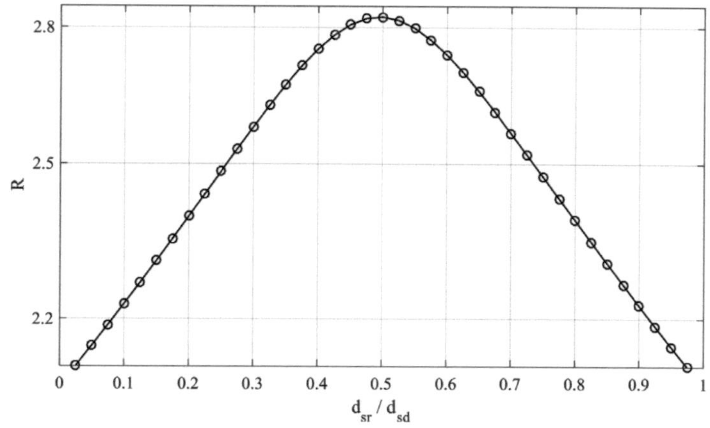

Figure 4. R vs. (d_{sr}/d_{sd}) for $(P_o/N_0) = 10$ dB.

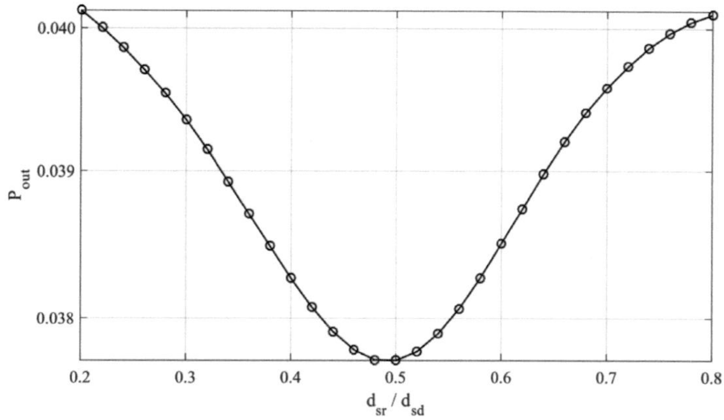

Figure 5. P_{out} for varying (d_{sr}/d_{sd}) with $(P_o/N_0) = 10$ dB.

Figure 5 depicts the outage probability performance as a function of $(\mathrm{d_{sr}}/\mathrm{d_{sd}})$. Here, $(\mathrm{P_o}/N_0) = 10$ dB, $\alpha = 4$, $\varepsilon_S = 0.1$, $\varepsilon_p = 0.05$, $\theta = \pi/2$ and $\mathrm{d_{sp}} = 2\,\mathrm{d_{sd}}$. The results obtained in **Figure 4** closely match with the results in **Figure 5**. Therefore, it can be deduced that the optimal placement of the relay terminal can be performed based on $(\mathrm{d_{sr}}/\mathrm{d_{sd}}) = 0.5$, which leads to the midpoint of the transmission link of $\mathrm{SU_s} - \mathrm{SU_d}$ as the optimal position.

In **Figure 6**, the transmission rate for the $\mathrm{PU_d} - \mathrm{SU_s}$ link is monitored for the normalized $\mathrm{d_{sd}}$ distance over Rayleigh fading channel while $(\mathrm{P_o}/N_0) = 10$ dB, $\alpha = 4$, $\varepsilon_S = 0.1$, $\varepsilon_p = 0.05$, $\theta = \pi/2$ and $\mathrm{d_{sr}} = \mathrm{d_{sd}}/2$. In addition, $\mathrm{P_{out}}$ performance analysis is also studied for the transmission link for $\mathrm{PU_d} - \mathrm{SU_s}$ with the normalized distance of $\mathrm{d_{sd}}$ and demonstrated in **Figure 7** using the same parameters in **Figure 6**.

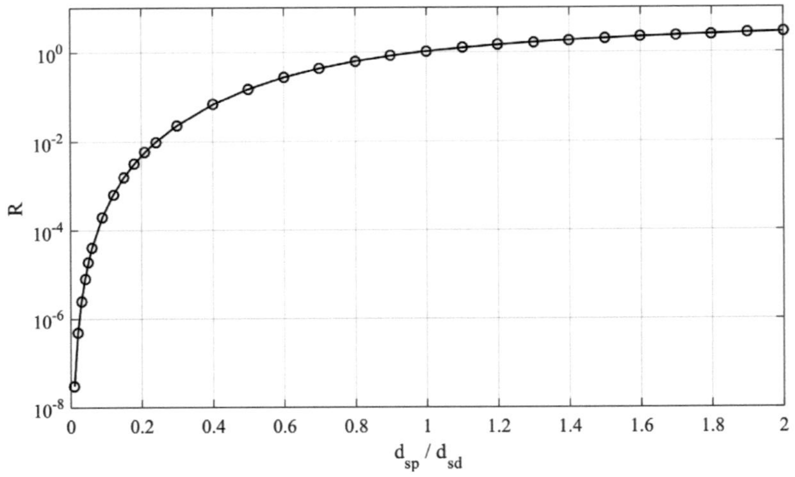

Figure 6. R vs. $\left(d_{sp}/d_{sd}\right)$ over Rayleigh fading channel while $(P_o/N_0) = 10$ dB.

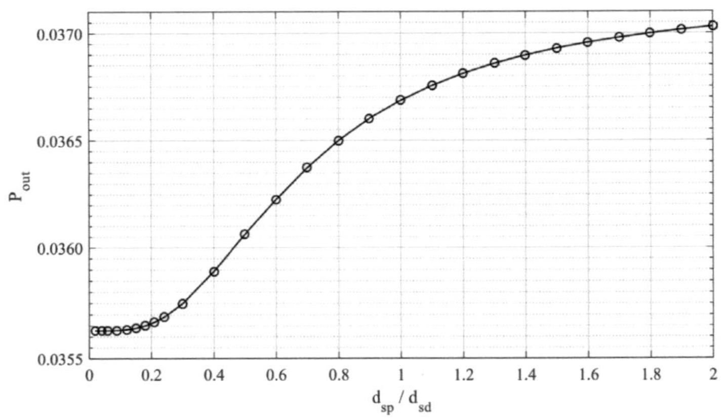

Figure 7. P_{out} performance with varying $\left(d_{sp}/d_{sd}\right)$ while $(P_o/N_0) = 10$ dB.

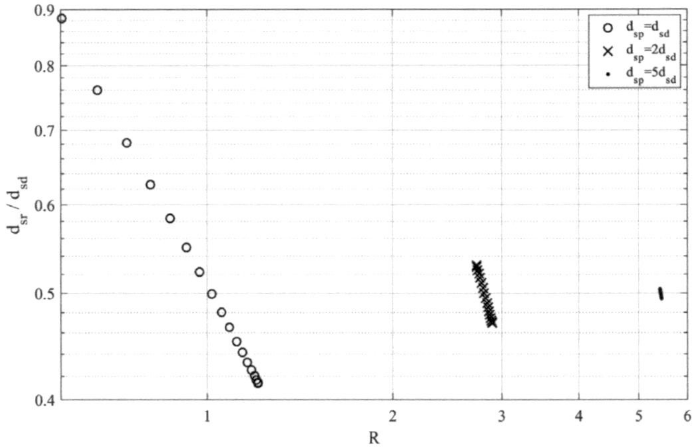

Figure 8. (d_{sr}/d_{sd}) vs. R over Rayleigh fading channel with different θ values for $(P_o/N_0) = 10$ dB, $d_{sp} = d_{sd}$, $d_{sp} = 2\,d_{sd}$ and $d_{sp} = 5\,d_{sd}$.

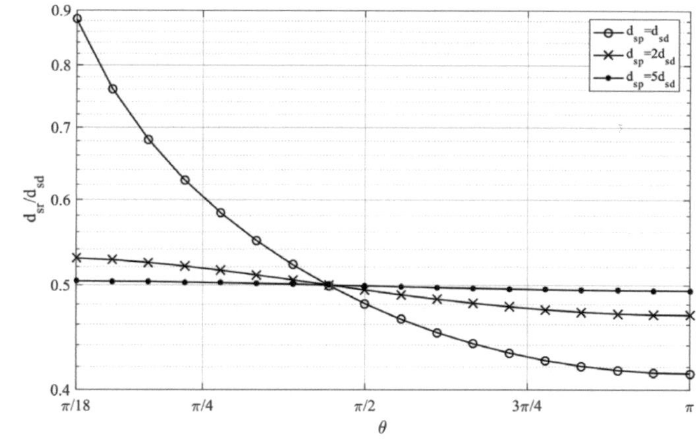

Figure 9. (d_{sr}/d_{sd}) vs. θ values for $d_{sp} = d_{sd}$, $d_{sp} = 2\ d_{sd}$ and $d_{sp} = 5\ d_{sd}$ while $(P_o/N_0) = 10$ dB.

The normalized d_{sr} distance varying with the transmission rate R over Rayleigh fading channel for different θ values and transmission links, $d_{sp} = d_{sd}$, $d_{sp} = 2\ d_{sd}$ and $d_{sp} = 5\ d_{sd}$ are shown in **Figure 8**. Besides, in **Figure 9**, d_{sr}/d_{sd} normalized distances are calculated for the different θ angles with varying d_{sp} values. Here, both figures are plotted for the values of $(P_o/N_0) = 10$ dB, $\alpha = 4$, $\varepsilon_S = 0.1$ and $\varepsilon_P = 0.05$.

The maximum transmission rate varying with different θ values for $d_{sp} = d_{sd}$, $d_{sp} = 2\ d_{sd}$ and $d_{sp} = 5\ d_{sd}$, while $(P_o/N_0) = 10$ dB is depicted in **Figure 10**. The figure demonstrates the effect of d_{sp} with varying θ angles. The results show that the maximum transmission rate of the considered system increases while θ and d_{sp} increases.

Finally, the maximum transmission rate, varying with the normalized distance for different d_{sp} values, is depicted in **Figure 11**. It is seen that while the d_{rp}/d_{sd} increases, the system performance also increases when θ is in the interval of $[0 - \pi]$. In other words, these results also prove that the R performance is directly related with the $PU_d - SU_s$ transmission link. While in case of d_{sp} distance is increased, the maximum transmission is achieved.

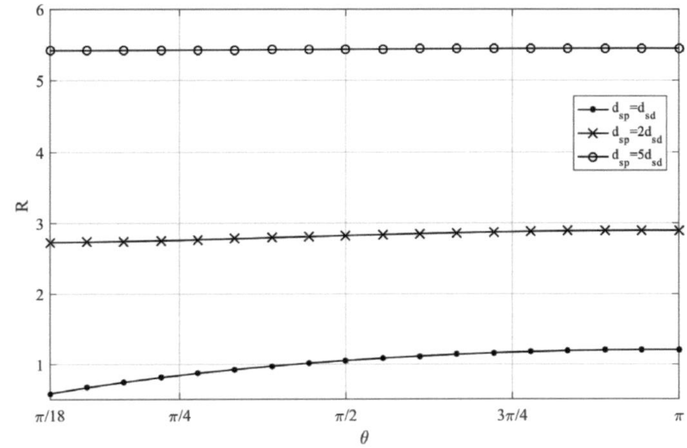

Figure 10. Maximum transmission rate varying with different θ values for $d_{sp} = d_{sd}$, $d_{sp} = 2\ d_{sd}$ and $d_{sp} = 5\ d_{sd}$ while $(P_o/N_0) = 10$ dB.

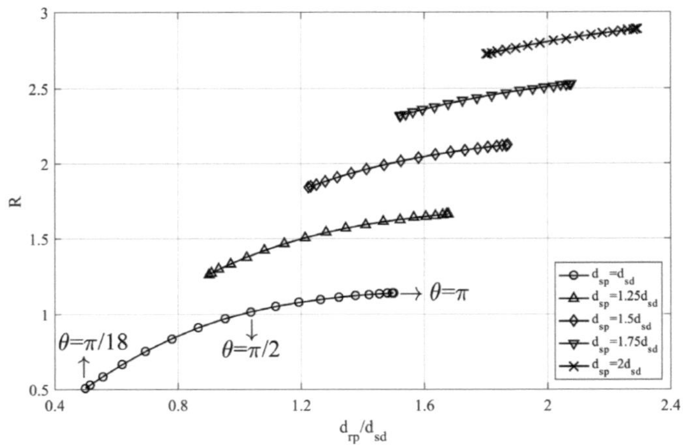

Figure 11. Maximum transmission rate varying with d_{rp} values normalized with d_{sd}, for different $PU_d - SU_s$ distance while $d_{sr} = d_{sd}/2$ and $(P_o/N_0) = 10$ dB.

4. Conclusions

In this chapter, we present a comprehensive performance analysis of the outage probability (P_{out}) and transmission rate (R) of the underlay cognitive radio networks with decode-and-forward relaying over Rayleigh fading channel. We provide a rigorous data for the optimal locations of the relay terminal using differential evolution optimization algorithm. We investigate the maximum transmission rate of the secondary user, and the outage probability subject to the distance of d_{sp}, d_{sr}, d_{rp}, normalized with d_{sd} between $PU_d - SU_s$, $SU_s - r$ and $PU_d - r$ transmission links, respectively. We then present the effect of the θ angle, between $PU_d - SU_s$ link and the horizontal axis, on the P_{out} and R performance. The numerical results, validates the theoretical analysis, show that d_{sp} distance and θ angle, which is in the interval of $[0 - \pi]$, have significant performance improvement on the transmission rate and the outage probability.

Acknowledgements

This work was supported in part by the Research Fund of Dumlupinar University under Scientific Research Project BAP/2016-84.

Author details

Mustafa Namdar* and Arif Basgumus

*Address all correspondence to: mustafa.namdar@gmail.com

Department of Electrical and Electronics Engineering, Dumlupinar University, Kutahya, Turkey

References

[1] Yucek T, Arslan H. A survey of spectrum sensing algorithms for cognitive radio applications. IEEE Communications Surveys & Tutorials. 2009;**11**(1):116–130. DOI: 10.1109/SURV.2009.090109

[2] Ma J, Li GY, Juang BH. Signal processing in cognitive radio. Proceedings of the IEEE. 2009;**97**(5):805–823. DOI: 10.1109/JPROC.2009.2015707

[3] Letaief KB, Zhang W. Cooperative communications for cognitive radio networks. Proceedings of the IEEE. 2009;**97**(5):878–893. DOI: 10.1109/JPROC.2009.2015716

[4] Akyildiz IF, Lee WY, Vuran MC, Mohanty S. NeXt generation/dynamic spectrum access/cognitive radio wireless networks: A survey. Computer Networks. 2006;**50**(13):2127–2159. DOI: 10.1016/j.comnet.2006.05.001

[5] Namdar M, Ilhan H, Durak-Ata L. Spectrum sensing for cognitive radio with selection combining receiver antenna diversity. In: IEEE 21st Signal Processing and Communications Applications Conference, April 2013; Girne, Northern Cyprus. pp. 1–4

[6] Namdar M, Sahin B, Ilhan H, Durak-Ata L. Chirp-z transform based spectrum sensing via energy detection. In: IEEE 20th Signal Processing and Communications Applications Conference, April 2012; Mugla, Turkey. pp. 1–4

[7] Namdar M, Ilhan H, Durak-Ata L. Partial spectrum utilization for energy detection in cognitive radio networks. In: IEEE International Congress on Ultra Modern Telecommunications and Control Systems, October 2012; St. Petersburg, Russia. pp. 989–994

[8] Namdar M, Ilhan H, and Durak-Ata L. Dispersed chirp-z transform-based spectrum sensing and utilization in cognitive radio networks. IET Signal Processing. 2014;**8**(4):320–329. DOI: 10.1049/iet-spr.2013.0127

[9] Namdar M, Ilhan H, and Durak-Ata L. Optimal detection thresholds in spectrum sensing with receiver diversity. Wireless Personal Communications, 2016;**87**(1):63–81. DOI: 10.1007/s11277-015-3026-6

[10] Sendonaris A, Erkip E, Aazhang B. User cooperation diversity—part I: System description. IEEE Transactions on Communications. 2003;**51**(11):1927–1938. DOI: 10.1109/TCOMM.2003.818096

[11] Sendonaris A, Erkip E, Aazhang B. User cooperation diversity—part II: Implementation aspects and performance analysis. IEEE Transactions on Communications. 2003;**51**(11):1939–1948. DOI: 10.1109/TCOMM.2003.819238

[12] Laneman JN, Wornell GW. Distributed space-time-coded protocols for exploiting cooperative diversity in wireless networks. IEEE Transactions on Information Theory. 2003;**49**(10):2415–2425. DOI: 10.1109/TIT.2003.817829

[13] Laneman JN, Tse DNC, Wornell GW. Cooperative diversity in wireless networks: Efficient protocols and outage behavior. IEEE Transactions on Information Theory. 2004;**50**(12):3062–3080. DOI: 10.1109/TIT.2004.838089

[14] Ilhan H. Performance analysis of cooperative vehicular systems with co-channel interference over cascaded Nakagami-m fading channels. Wireless Personal Communications. 2015;**83**:203–214. DOI: 10.1007/s11277-015-2389-z

[15] Basgumus A, Hicdurmaz B, Temurtas H, Namdar M, Altuncu A, Yilmaz G. Optimum transmission distance for relay-assisted free-space optical communication systems. Elsevier Optik. 2016;**127**(16):6490–6497. DOI: 10.1016/j.ijleo.2016.04.070

[16] Basgumus A, Namdar M, Yilmaz G, Altuncu A. Performance comparison of the differential evolution and particle swarm optimization algorithms in free-space optical communications systems. Advances in Electrical and Computer Engineering. 2015;**15**(2):17–22. DOI: 10.4316/AECE.2015.02003

[17] Basgumus A, Namdar M, Yilmaz G, Altuncu A. Performance analysis of the differential evolution and particle swarm optimization algorithms in cooperative wireless communications. In: Baskan O. editor. Optimization Algorithms-Methods and Applications. Rijeka, Croatia: InTech. 2016. ISBN: 978-953-51-2593-8. DOI: 10.5772/62453

[18] Nasri A, Schober R, Blake IF. Performance and optimization of cooperative diversity systems in generic noise and interference. In: IEEE International Communications Conference. May 2010; Cape Town, South Africa. pp. 1132–1143

[19] Woong C, Liuqing Y. Optimum resource allocation for relay networks with differential modulation. IEEE Transactions on Communicaitons. 2008;**56**(4):531–534. DOI: 10.1109/TCOMM.2008.060104

[20] Mohammed H, Khalaf TA. Optimal positioning of relay node in wireless cooperative communication networks. In: IEEE 9th International Computer Engineering Conference, December 2013; pp. 122–127

[21] Lloyd E, Xue G. Relay node placement in wireless sensor networks. IEEE Transactions on Computers. 2007;**56**(1):134–138. DOI: 10.1109/TC.2007.250629

[22] Han B, Li J, Su J. Optimal relay node placement for multi-pair cooperative communication in wireless networks. In: IEEE Wireless Communications and Networking Conference, April 2013; San Francisco, USA. pp. 4724–4729

[23] Yue H, Pan M, Fang Y. Glisic S. Spectrum and energy efficient relay station in placement in cognitive radio networks. IEEE Journal on Selected Areas in Communications. 2013;**31**(5):883–893. DOI: 10.1109/JSAC.2013.130507

[24] Palombara CL, Tralli V, Masini BM, Conti A. Relay-assisted diversity communications. IEEE Transactions on Vehicular Technology. 2013;**62**(1):415–421. DOI: 10.1109/TVT.2012.2218841

[25] Ikki SS, Aissa S. Study of optimization problem for amplify-and-forward relaying over Weibull fading channels. In: IEEE 72nd Vehicular Technology Conference Fall, September 2010; Ottowa, Canada. pp. 1–5

[26] Han L, Mu J, Wang W, Zhang B. Optimization of relay placement and power allocation for decode-and-forward cooperative relaying over correlated shadowed fading channels.

EURASIP Journal on Wireless Communications and Networking. 2014;**2014**(41):1–7. DOI: 10.1186/1687-1499-2014-41

[27] Zhu J, Huang J, Zhang W. Optimal one-dimensional relay placement in cognitive radio networks. In: IEEE International Conference on Wireless Communications & Signal Processing, October 2010; Suzhou, China. pp. 1–6

Section 3

Hardware Aspects

Analog-to-Digital Conversion for Cognitive Radio: Subsampling, Interleaving, and Compressive Sensing

José Ramón García Oya,
Fernando Muñoz Chavero and
Rubén Martín Clemente

Additional information is available at the end of the chapter

http://dx.doi.org/10.5772/intechopen.69241

Abstract

This chapter explores different analog-to-digital conversion techniques that are suitable to be implemented in cognitive radio receivers. This chapter details the fundamentals, advantages, and drawbacks of three promising techniques: subsampling, interleaving, and compressive sensing. Due to their major maturity, subsampling- and interleaving-based systems are described in further detail, whereas compressive sensing-based systems are described as a complement of the previous techniques for underutilized spectrum applications. The feasibility of these techniques as part of software-defined radio, multistandard, and spectrum sensing receivers is demonstrated by proposing different architectures with reduced complexity at circuit level, depending on the application requirements. Additionally, the chapter proposes different solutions to integrate the advantages of these techniques in a unique analog-to-digital conversion process.

Keywords: analog-to-digital conversion, cognitive radio, compressive sensing, interleaving, multistandard receivers, software-defined radio, spectrum sensing, subsampling

1. Introduction

Analog-to-digital conversion (ADC) stage is one of the main bottlenecks of the high-speed telecommunications systems. This chapter presents a survey of different feasible analog-to-digital conversion techniques that are suitable to overcome these difficulties and to get the software-defined radio (SDR) paradigm [1], where most functionalities, instead of being performed in the analog domain (i.e., filters and mixers), are performed in the digital domain. In SDR, the analog-to-digital conversion is implemented immediately after the antenna, and the radio frequency (RF) signal is directly converted to digital without any previous mixing stage. Since it is not possible to approach this idea with traditional analog-to-digital conversion from current commercial devices, this chapter describes some techniques that may be employed instead. Although the proposed systems have more restrictive specifications, these solutions reduce the final complexity, as will be detailed in this chapter. This work explores three different promising techniques: subsampling, interleaving, and compressive sensing (CS).

The proposed techniques are an appealing solution to approach the cognitive radio (CR) objectives [2, 3], which are conditioned by the physical implementation of the SDR receiver. Due to the necessity of several wireless standards coexisting in the same device, a high flexibility and programmability will be an important requirement for the proposed architectures, with the objective of being employed in multistandard receivers.

With these objectives in mind, this chapter describes in detail three different approaches for implementing the analog-to-digital conversion stage. The choice of one or other of these approaches will depend on the environment, the properties of the received signals, and the parameters that have to be optimized. For receivers where the main requirements are a high-resolution and a high-analog bandwidth that covers a maximum number of communication standards, we propose a system based on subsampling techniques. For applications where the main requirement is to maximize the data acquisition rate, the proposed system is based on interleaving techniques, that is, the interconnection of several analog-to-digital converters in parallel. Finally, compressive sensing techniques will be preferred for scenarios where the spectrum can be considered sparse, that is, for a wideband spectrum with a low spectral occupancy, where it will be possible to recover the received signal and implement an estimation of the radio channel by using a reduced number of samples from the ADC. This emerging technology will be presented at architectural level, so that it will be studied from the point of view of its integration with the two main techniques detailed in this chapter, that is, subsampling and interleaving techniques, with the objective of exploiting their advantages for sparse spectrum sensing applications.

2. Subsampling techniques

2.1. Overview of the subsampling concepts

2.1.1. Subsampling idea

Subsampling-based systems, whose conceptual diagram is illustrated in **Figure 1**, are a feasible alternative to the classic mixing-based receiver architectures. The received signal is filtered by an RF band-pass filter that can be a tunable filter or a bank of filters. The incoming band-pass signal is sampled below the Nyquist rate [4, 5], being possible to avoid aliasing using some sampling properties. This sampled signal is converted to digital using an ADC at intermediate sampling rate. The main advantage of this scheme is its simplicity, because the number of components is reduced, being possible to place the data conversion very close to the antenna. As a consequence, many operations like filtering, frequency translation, and demodulation can

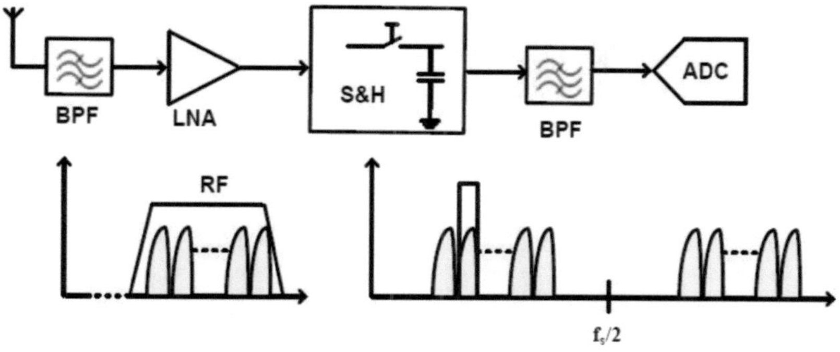

Figure 1. Conceptual diagram of the subsampling receiver architecture.

be implemented in digital domain, taking advantage of low-cost digital very large-scale integration (VLSI) solutions, and leading to a high integration while eliminating problems such as DC offset and 1/f noise.

In addition to system cost reduction, pushing these functions in digital domain eliminates many of the sensitivities of analog solutions, such as device matching, environmental sensitivity, and performance variation over time. The flexibility and reconfigurability required by SDR applications are also increased by moving the ADC into IF stage and, moreover, it is possible to use this architecture for wideband and multistandard applications because of its large analog bandwidth. In this architecture, a single ADC can sample multiple signal channels, which are then separated and demodulated in parallel in digital domain.

In **Figure 1**, the Sample & Hold (S&H) after the low noise amplifier (LNA) downconverts the RF signal to intermediate frequency. This enables us to relax the constraints on the S&H inside the ADC: if the bandwidth of the ADC is required to include the RF carrier frequency, it may not be possible to simultaneously fulfill the conditions on the required dynamic range and resolution of the ADC using the current technologies. To mention other problems, subsampling receivers have several noise sources, such as the jitter and thermal noise folded in the band of interest, which will be detailed in the subsequent sections.

Finally, RF band-pass filtering is required when avoiding overlap between folded signals is necessary. These BP filters, especially on-chip filters, are difficult to implement at high frequencies. Although external filters, such as SAW filters, can be used, they are only available at limited number of frequencies, so this is not a practical solution to design multistandard receivers. Alternatively, increasing the sampling frequency can decrease the selectivity of the filter. However, this solution has some drawbacks, such as the high technology and high cost required by the ADC, whose resolution and dynamic range will be degraded as compared to lower sample rate ADC alternatives. Also, power consumption increases with sample rate. Therefore, the cost, performance and power consumption of other devices (such as ADC clock sources or digital circuits after the ADC) also depend on the ADC sample rate. In this section, we address the problem of planning the sampling frequencies with the objective of avoiding this overlapping between signals and reduce the complexity of the RF filtering. Note that we do not consider additional adjacent interferers not overlapped with the desired signal since they can be suppressed by additional channel filtering in digital domain.

2.1.2. Selecting the optimal sampling frequency

For an input signal with carrier frequency f_c and analog bandwidth BW, the minimal sampling frequency f_s is established by the Nyquist theorem, $f_s > 2B$, being $B = f_c + BW/2$. However, if we are processing band-pass signals, aliasing can be avoided with a lower f_s when Eq. (1) is satisfied [5]:

$$\frac{2f_c - BW}{m-1} > f_s > \frac{2f_c + BW}{m} \tag{1}$$

where m is an integer number representing the number of replicas of the original signal that appears in the range $[0, f_c - BW/2]$. Note also that f_s has to be higher than two times BW. The maximum number of copies needed to avoid aliasing is calculated by Eq. (2) [5]:

$$m_{\max} = floor((f_c + BW/2)/BW) \tag{2}$$

A suitable value in the range given by Eq. (1) is the one that produces a copy on $f_s/4$. This frequency is given by Eq. (3) [5]:

$$f_s = \frac{4f_c}{m_{\text{odd}}} \tag{3}$$

where m_{odd} is an integer odd number greater than 1. Moreover, if this value is 5, 9, 13,…, there is no spectral inversion. Sampling at $4f_c/m_{\text{odd}}$ results in a larger subsampling frequency bandwidth and relaxes the filtering requirements after the S&H.

2.1.3. Subsampling nonidealities

This section is centered in the main nonidealities of the S&H illustrated in **Figure 1**, which is the most critical device in the subsampling-based systems, as it processes high-frequency signals.

2.1.3.1. Jitter noise

Ideally, the time interval between samples is a constant value equal to $1/f_s$. Nevertheless, in practice there are fluctuations due to jitter. This jitter produces an increment of the output total noise, thus limiting the effective number of bits (ENOB). Jitter is produced by two different sources: the phase noise associated to the oscillator and the aperture jitter of the S&H. At a first approach, we can consider both sources of jitter as noncorrelated Gaussian stochastic processes. When the input is a sinusoidal signal like $y(t) = A\sin(2\pi f_{\text{in}} t)$, signal-to-noise ratio (SNR) at the S&H output is determined by [6, 7]

$$\mathrm{SNR} = \frac{\frac{A^2}{2}}{\overline{N}_\tau} = \begin{cases} \dfrac{1}{4\pi^2 f_{\mathrm{in}}^2 \sigma_\tau^2} : 2\pi f_{\mathrm{in}} \sigma_\tau << 1 \\ \dfrac{1}{2(1 - e^{-2\pi^2 f_{\mathrm{in}}^2 \sigma_\tau^2})} : \text{otherwise} \end{cases} \tag{4}$$

where $\overline{N}_\tau$ is the average noise power and σ_τ is the jitter standard deviation. Note that the SNR is degraded when the input frequency increases.

2.1.3.2. Folded thermal noise

The S&H can be modeled as illustrated in **Figure 2a** [6], where the switch introduces a thermal noise with a power spectral density (PSD) equal to $S_{\mathrm{in}}(f) = 2kTR_{\mathrm{on}}$, where k is the Boltzmann constant, T is the temperature, and R_{on} is the on-resistance of the switch. This noise is AWGN (additive white Gaussian noise) and is folded in the band of interest by the subsampling process (**Figure 2b**).

R_{on} and C model a low-pass (LP) filter (**Figure 2a**) with transfer function $H(f) = 1/(1 + j2\pi f R_{\mathrm{on}} C)$, whose 3-dB cutoff frequency is equal to $f_{3\mathrm{dB}} = 1/(2\pi R_{\mathrm{on}} C)$. The output PSD will be [7]

$$S_{\mathrm{out}}(f) = S_{in}(f)|H(f)|^2 = 2kTR_{\mathrm{on}} \frac{1}{1 + 4\pi^2 f^2 R_{\mathrm{on}}^2 C^2} \tag{5}$$

Being the total noise power (by a two-sided representation):

$$P_{\mathrm{out}} = \int_{-\infty}^{\infty} S_{\mathrm{out}}(f) df = \frac{kT}{C} \tag{6}$$

For modeling purposes, the output noise can be considered to be a Gaussian thermal noise filtered by a brick-wall filter of bandwidth equal to B_{eff} (i.e., the noise bandwidth in **Figure 2c**) [7]:

$$B_{\mathrm{eff}} = \frac{1}{4R_{\mathrm{on}} C} = \frac{\pi}{2} f_{3\mathrm{dB}} \tag{7}$$

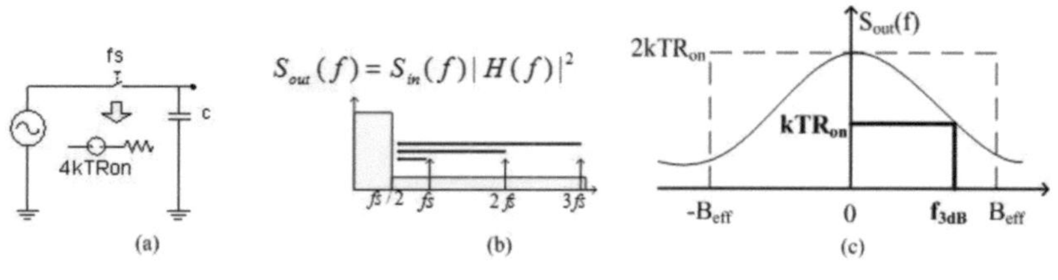

Figure 2. (a) S&H model, (b) noise folded in the band of interest, and (c) effective noise bandwidth.

Therefore, the noise power can be rewritten as follows [7]:

$$P_{out} = \frac{kT}{C} = 2kTR_{on} \cdot (2B_{eff}) \tag{8}$$

On the other hand, the SNR in $[-B_{eff}, B_{eff}]$ is defined as [6]

$$SNR = \frac{P_s}{N_i + (m-1)N_o} \tag{9}$$

where P_s is the signal power spectral density, m is the number of replicas, and N_i and N_o are the in-band and the out-of-band noise power spectral densities, respectively. Setting $2B_{eff} = mf_s$, when $m = 1$ the Nyquist theorem is met and the SNR is not affected by the folded noise. Otherwise, if $m > 1$, and assuming $N_i = N_o = N$:

$$SNR = \frac{P_s}{mN} = \frac{P_s}{N(2B_{eff}/f_s)} \tag{10}$$

Thereby, the out-of-band folded noise reduces the SNR by a factor $2B_{eff}/f_s$. From Eq. (10), we can observe that the SNR increases around 3 dB when the sampling frequency is doubled. Thus, regarding the noise, it is convenient to select the largest sampling frequency among the set of possible sampling frequencies defined by the digital signal processing block specifications.

2.2. Noise performance optimization: multiple clocking techniques

This section describes a method for improving the resolution that employs two consecutive subsampling stages, with the objective to reduce the folded noise effect. The use of two subsampling processes enables us to increase the sampling frequency of the first stage, resulting in a lower contribution of the first S&H to the folded thermal noise.

Figure 3 illustrates two different alternatives to implement a subsampling-based receiver. **Figure 3a** shows the scheme for a unique clock, whereas **Figure 3b** shows the scheme with two different clocks. The scheme illustrated in **Figure 3a** was employed in Ref. [8] by using the S&H Inphi 1821TH and the ADC E2V AT84AS001, with a maximum sampling frequency of 2 GHz and 500 MHz, respectively. Therefore, the unique sampling frequency for this architecture is limited to 500 MHz. However, the sampling frequency of the first S&H in **Figure 3b** can be selected up to 2 GHz, obtaining a band-limited signal at the output [9]. As the first sampling frequency is very large, the folded thermal noise added by this stage is reduced. After filtering, the resulting signal is sampled again by a second S&H at 400–500 MHz.

Figure 3 also illustrates the folded noise (dotted line) for the single clock (**Figure 3a**) and the multiple clock (**Figure 3b**) cases, considering for both cases the same thermal noise level at the input of the S&H (solid line) and from the ADC (dashed line). Since the effective noise bandwidth of the S&H is typically much larger than that of the ADC, the improvement achieved at the S&H in **Figure 3b** is usually dominant. In **Figure 3b**, a BP filter is necessary to select a proper band and decrease the out-of-band noise folded by the second subsampling process, while an LP filter with a cutoff frequency equal to $f_s/2$ may be enough in **Figure 3a**.

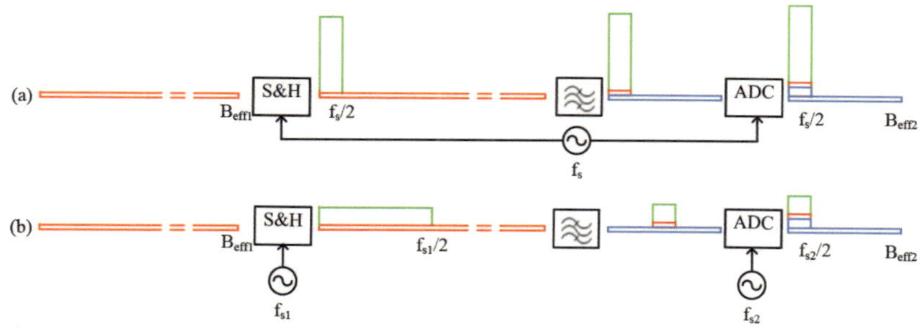

Figure 3. Folded noise effects using single clock (a) and multiple clock (b).

Assuming that the noises of both S&Hs in **Figure 3** are uncorrelated, from Eq. (10), the output power spectral density due to the S&Hs white noise in **Figure 3a** and b, respectively, is [9] as follows:

$$P_{N(a)} = \frac{2B_{\text{eff1}}}{f_s}N_1 + \frac{2B_{\text{eff2}}}{f_s}N_2\,;P_{N(b)} = \frac{2B_{\text{eff1}}}{f_{s1}}N_1 + \frac{2B_{\text{eff2}}}{f_{s2}}N_2 \tag{11}$$

Where N_1 and N_2 are the noise power introduced by $S\&H_1$ and $S\&H_2$, respectively, and B_{eff1} and B_{eff2} their respective noise bandwidths. Note that there will not be folding of N_1 during the second sampling process because the signal is filtered at IF in both cases using a BP filter. Therefore, the SNR improvement obtained with this multiple clocking method is given by [9]

$$\frac{SNR_{(b)}}{SNR_{(a)}} = \frac{P_s/N_{(b)}}{P_s/N_{(a)}} = \frac{B_{\text{eff1}}N_1 + B_{\text{eff2}}N_2}{\frac{f_s}{f_{s1}}B_{\text{eff1}}N_1 + \frac{f_s}{f_{s2}}B_{\text{eff2}}N_2} \tag{12}$$

As the first S&H processes high-frequency signals, $B_{\text{eff1}} >> B_{\text{eff2}}$. In addition, we can consider the noise power spectral densities of both S&Hs with the same order of magnitude. Then, Eq. (12) can be approximated by [9]

$$\frac{SNR_{(b)}}{SNR_{(a)}} \approx \frac{1}{\frac{f_s}{f_{s1}} + \frac{f_s}{f_{s2}}\frac{B_{\text{eff2}}}{B_{\text{eff1}}}} \tag{13}$$

The most influential term in this improvement is f_{s1}/f_s, that is, a higher value of this ratio will mean a better SNR improvement. The experimental results obtained for an analog input signal bandwidth of 20 MHz are given in **Figure 4**, where it is possible to observe the improvement over the ENOB by using the proposed technique.

2.3. Subsampling-based systems for cognitive radio applications (I): approach for nonlinear and multi-band environments

2.3.1. Studied scenarios

The benefit of using a subsampling-based receiver for cognitive radio is, besides its simplicity and reconfigurability, that the capability of hopping between different spectrum spans only requires

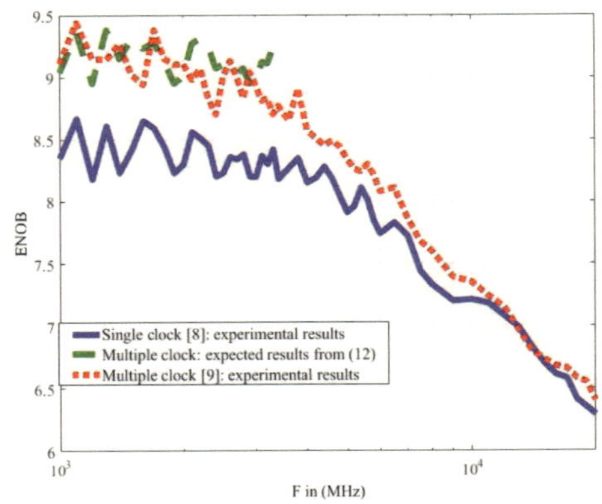

Figure 4. ENOB obtained using single clocking and multiple clocking techniques.

adjusting the sampling frequencies and selecting the appropriate band-pass filters. However, there is a challenge when using subsampling concurrently in a multi-signal environment and/or nonlinear conditions, because the replicas of the generated harmonics are folded back in the band of interest and may overlap with the desired signals. This issue was addressed in Ref. [10] where a universal formula for subsampling in nonlinear systems was developed for single-band applications. In dual-band receiver applications, the main problem of subsampling is the possible overlapping between the replicas of the two desired signals in the IF frequency band. This problem was studied in Ref. [11] for multi-band linear and noninterfering environments.

As one example of application, a dual-band subsampling receiver has been proposed for use in a feedback loop of a dual-band transmitter for linearization purposes using digital predistortion [12]. In Ref. [13], a subsampling receiver for dual-band applications was proposed, due its simplicity, to implement a cognitive radio sensing different bands and checking if they are in use. In Ref. [13], the designed subsampling receiver does not consider any interferers, harmonics, or intermodulation effects.

This section extends the above study [12, 13] by optimizing the SNR of concurrent dual-band signals at the receiver in a multi-signal or nonlinear environment. The requirement of increasing the analog bandwidth and reducing the effect of the folded noise leads to propose new receiver topologies with the objective of improving these features for a larger number of communication standards. Interferences and spurious signals in the received spectrum can be treated as intermodulation products using the same optimization technique that will be described later, so when these signals are subsampled the resulting aliasing components with these unwanted and spurious signals do not overlap with the desired signal.

As an additional benefit, these extra conditions used in the sampling frequency selection can lead to more relaxed RF filter requirements, since the known unwanted signals in the spectrum will not affect the desired signal bandwidth at IF and, therefore, they will be filtered more easily after being subsampled.

Figure 5 illustrates a basic scheme for a concurrent dual-band subsampling-based receiver. It consists of LNA, S&H, ADC, and band-pass filters (located in different parts of the receiver chain to filter out unwanted signals).

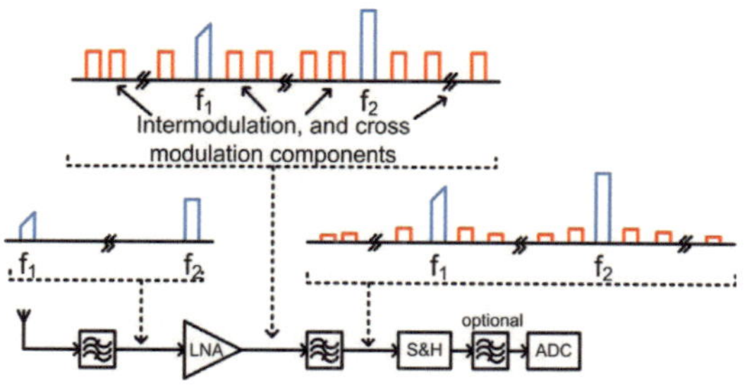

Figure 5. Subsampling receiver in multi-band nonlinear environment.

In concurrent dual-band applications, the LNA may introduce intermodulation and cross-modulation components at the receiver path right before the S&H block. Proper filtering at the receiver may attenuate the level of unwanted intermodulation components, without completely removing them all. In subsampling-based receivers, these intermodulation and cross-modulation leakage signals could be a significant source of noise for the in-band signals. In fact, it depends on the selected subsampling frequency, and proper and careful subsampling frequency selection could avoid the overlapping between the desired in-band signals and the unwanted intermodulation and cross-modulation signals leakage through the band-pass filter. **Figure 5** also illustrates the signal spectra at different points of the receiver. The purpose of the first band-pass filter is to remove the out-band undesired spectrum. Because of the nonlinear nature of the concurrent dual-band LNA, intermodulation and cross-modulation components are generated in the receiver path. The second band-pass filter attenuates these unwanted components as much as possible. The signals at the input of S&H are the two desired signals plus those unwanted spurious components with a signal level higher than the noise floor.

The signals produced by the LNA nonlinearity can be overlapped when subsampled by the S&H. The architectures proposed in this work to avoid this overlapping are based on single and multiple clock techniques, where the objective is to optimize the noise performance and the flexibility of the system for its use in multistandard applications. The work presented in this section encompasses an analysis of the multi-signal subsampling receiver from noise and nonlinear distortion perspectives, and proposes optimized architectures to mitigate these aspects and to improve the overall performance of the subsampling receiver in multi-signal environment in terms of signal quality and subsampling speed, along with an experimental validation.

2.3.2. Subsampling in nonlinear environments

When an input signal centered at f_1 (**Figure 6a**) [10] drives a nonlinear system, the output signal may produce multiple spectra, with different bandwidths, centered at integer multiples of f_1 (see **Figure 6a**). Let us consider any two intermodulation products at frequencies if_1 and jf_1, with respective bandwidths B_i and B_j and $j > i$. It can be easily shown (see Ref. [10]) that the range of the sampling frequencies that guarantees that those intermodulation products do not overlap in the sampled output spectrum is given by

$$\frac{kf_1 + W}{n+1} < f_s \leq \frac{kf_1 - W}{n} \tag{14}$$

where $k = j - i$, $W = (B_i + B_j)/2$ and $n = \text{floor}((jf1 - if1)/fs)$.

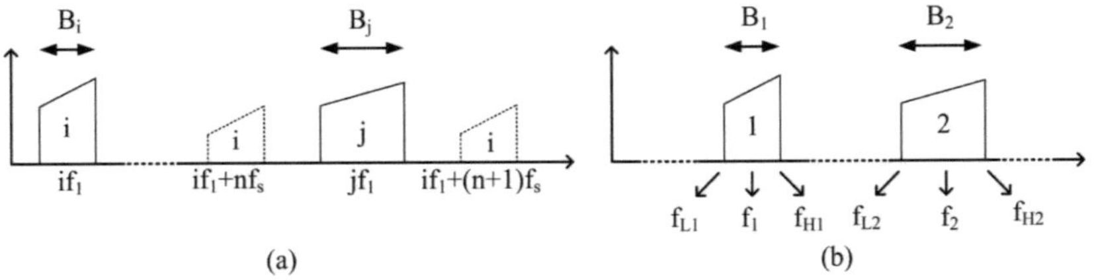

Figure 6. Frequency locations for nonlinear (a) and dual-band (b) scenarios.

2.3.3. *Subsampling for multi-band systems*

An algorithm to find the range of valid subsampling frequencies for multi-band systems is presented in Ref. [11]. For the particular case of a dual-band input spectrum (**Figure 6b**), the subsampling frequency f_s must be chosen to ensure that the two signals do not overlap in the subsampled domain. From the general equations presented in Ref. [11] and considering a dual-band case, it follows that f_s satisfies the following equation:

$$n_1 = \left\lfloor \frac{f_{L1}}{f_s} \right\rfloor \leq \left\lfloor \frac{f_{L1}}{2((f_{U1} - f_{L1}) + (f_{U2} - f_{L2}))} \right\rfloor \tag{15}$$

where $f_{L1}$and f_{U1} are the lower and upper limits of the lower band and f_{L2} and f_{U2} are lower and upper limits of the upper band, being n_1 the maximum replica order of the lower band. Denoting $R_1 = f_2/f_1$, we have that f_s must also verify [11]:

$$\lfloor R_1 n_1 \rfloor \leq n_2 \leq \lfloor R_1 n_1 + R_1 \rfloor \tag{16}$$

where $n_2 = \text{floor}(f_2/f_s)$.

2.3.4. *Subsampling for multi-band systems in nonlinear environments*

This section describes the algorithm employed to integrate both scenarios previously described, that is, multi-band and nonlinear systems which utilize subsampling techniques. The final sampling ranges will be given by the following expression:

$$F = F_{\text{db}} \cap F_{\text{imd}} \cap F_{\text{cmd}} \cap F_{\text{hmd}} \tag{17}$$

where F is the intersection of all the valid ranges calculated from Eqs. (14)–(16), being F_{db}, F_{imd}, F_{cmd}, and F_{hmd} are the valid sampling frequency sets for the fundamental signals, intermodulation, cross-modulation, and harmonic distortion, respectively.

In order to find *F*, an algorithm has been developed in MATLAB [14]. This algorithm calculates these ranges and the location of the replicas, where the input parameters are the signal frequencies, the signal bandwidths, and the number of harmonics (i.e., order of the nonlinearity). Therefore, knowing the signal frequencies and the number of harmonics, it is possible to calculate the location of the intermodulation and the cross-modulation products. An example of the results obtained from this algorithm is illustrated in **Table 1** [14], showing the three first valid ranges immediately lower than 2 GHz for the input signals at 1.82 and 2.4 GHz, considering five harmonics and a signal bandwidth of 25 MHz. The subsampled spectrum is illustrated in **Figure 7** [14] for a sampling frequency equal to 2 GHz, showing that there is no overlapping between signals.

2.3.5. Optimization of dual-band receivers in nonlinear environments

An analysis focused on optimizing the noise performance is presented in this section. Independent clocks for the S&H and ADC are proposed to limit the noise effects, and a bank of band-pass filters is used to filter out most of the aliased nonlinear and interfering components. Several different subsampling architectures and filter configurations are analyzed in theoretical and measurement environments.

Lower frequency bound (MHz)	Upper frequency bound (MHz)
1995	2000
1837.5	1978.33
1801.67	1802.5

Table 1. Valid sampling frequencies below 2 GHz.

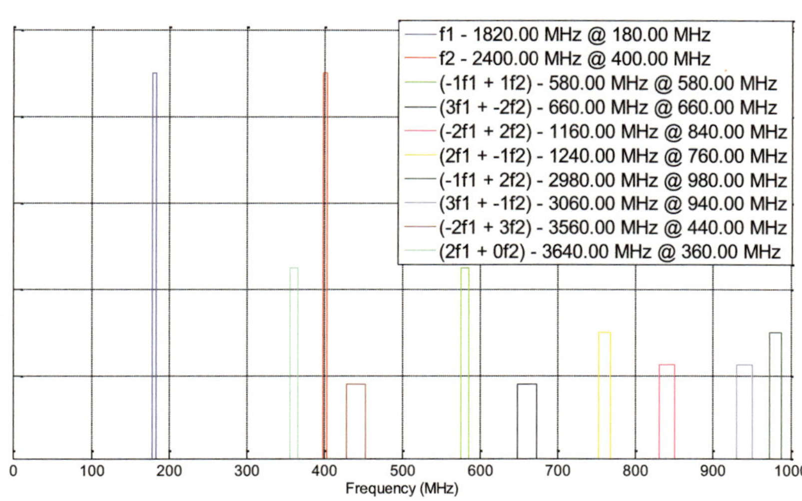

Figure 7. Subsampled spectrum for 1.82 and 2.4 GHz input frequencies.

2.3.5.1. Subsampling for concurrent dual-band and nonlinear systems using multiple clocking techniques

This section is focused on an optimized concurrent dual-band subsampling receiver for noise performance and versatility, in order to cover most wireless communication standards. The optimization takes advantage of the flexibility of subsampling, as it is possible to study different valid alternatives to clock the receiver in order to maximize its noise performance.

As described in Section 2.2, clocking the S&H and the ADC with the same clock limits the maximum sampling frequency of the system to that of the ADC, which is usually significantly lower than the maximum sampling frequency of the S&H. As a better alternative, employing an additional higher frequency to clock the S&H it is possible to increase the SNR of the receiver. Moreover, for this dual-band application, additional degrees of freedom can be obtained using a multiple clock scheme, so that we can cover a higher number of dual-band combinations of wireless communication standards. However, the BP filter used in **Figure 3b** might reduce the flexibility of the receiver when it is used in multi-band applications and in a nonlinear environment. This section tries to find the optimal filter bandwidth that reduces the folded noise, while avoiding a significant reduction in the number of valid sampling frequencies.

Figure 8 shows the effects of a third-order nonlinearity when a dual-band signal passes through a nonlinear subsampling receiver. In the first scenario (**Figure 8a**), the S&H and the ADC are clocked at the same rate. In the second scenario (**Figure 8b**), the clock rates of S&H and ADC are different. Both carrier frequencies at 880 MHz and 1.82 GHz are sampled at 400 MHz, where this frequency has been calculated by the algorithm described in Section 2.3.4. Using this sampling frequency, **Figure 8a** shows the different Nyquist bands along which the input signals and their harmonics and intermodulation products are distributed at the S&H input. All of them are folded to IF. After subsampling, the signal is filtered and converted to digital domain, where the Nyquist theorem is met. Similarly, **Figure 8b** presents the scenario where the S&H and ADC use different clock rates. In order to reduce the folded noise effect, the S&H sampling frequency is increased to 2 GHz, while the input RF signal at the S&H is the same as in the case of **Figure 8a**. The second sampling frequency is still 400 MHz. Both frequencies also satisfy the criteria detailed in the previous sections, so that aliasing between the target signals and their harmonics and intermodulation products is avoided. Finally, it is observed in **Figure 8b** that it is possible to relax the specifications of the filters, as desired, because of a higher first-sampling frequency

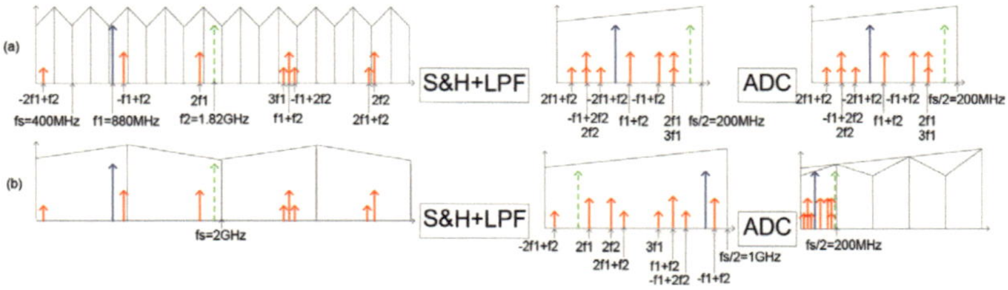

Figure 8. Folded effects for harmonics and intermodulation products using a single clock (a) and multiple clock (b) techniques.

places the IF components more separately in the spectrum, while maintaining at the same time the objective of removing the maximum number of the undesired components.

2.3.5.2. Experimental results

An experimental validation using realistic wireless communication scenarios is presented below to demonstrate the robustness and appropriateness of proposed technique in multi-standard environments. The seven input frequencies chosen to study the selective combinations for different dual-band applications correspond to the standards studied in Ref. [14]; the grouping for two given bands is shown in **Table 2**.

The experimental setup is illustrated in **Figure 9** [15]. The dual-band signals are continuous wave signals to demonstrate the peak SNR that can be achieved, and each signal band is generated by independent Agilent PSG E8257D signal generators. A power combiner combines both signal sources into a dual-band signal is amplified using a LNA ZX60-6013 from Minicircuits, and subsampled by an S&H Inphi 1821TH, and a ADS5474 ADC from Texas Instruments. The signal generators used as clock sources for the S&H and the ADC are the Agilent E8663D and the Rohde & Schwarz SMIQ. The implemented receiver architecture's design is shown in **Table 3**.

	Dual-band signal scenarios											
Standard	**1**	**2**	**3**	**4**	**5**	**6**	**7**	**8**	**9**	**10**	**11**	**12**
WCDMA (V) 880 MHz	x	x	x	x	x	x						
GSM-DCS 1.82 GHz	x						x	x	x	x	x	
WCDMA (I) 2.12 GHz		x					x					x
Bluetooth 4 GHz			x					x				x
WiMAX 3.5 GHz				x					x			
802.11a 5.2 GHz					x					x		
WiMAX 5.8 GHz						x					x	

Table 2. Dual-band signal construction table.

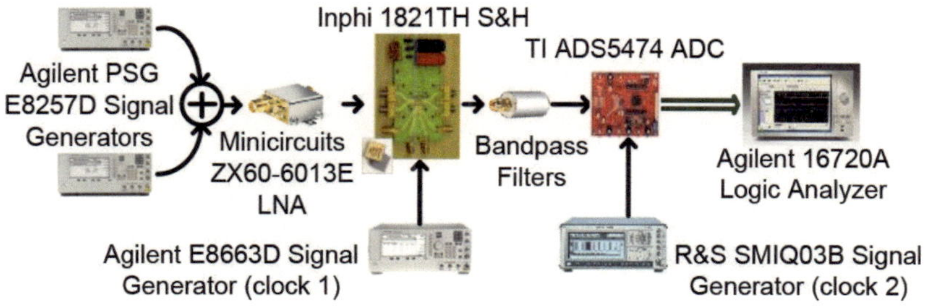

Figure 9. Experimental setup for dual-band subsampling receiver.

The simulated subsampled spectra for RF signals at 2.12 and 2.4 GHz (dual-band scenario 12, Receiver Design 4) are shown in **Figure 10**, where the expected bandwidth of each dual-band signal is 5 MHz, and we assume a fifth nonlinearity order. The S&H subsampling frequency was set to 1900 MHz, and the ADC subsampling frequency was set to 400 MHz. **Figure 11** shows the

Receiver design (RD)	Receiver's architecture
2	$f_{s1} = f_{s2} < 400$ MHz, without bank of filters
4	$f_{s1} < 2$ GHz, $f_{s2} < 400$ MHz, without bank of filters
5	$f_{s1} < 2$ GHz, $f_{s2} < 400$ MHz, with a bank of two filters in [0–400] and [400–800] MHz
6	$f_{s1} < 2$ GHz, $f_{s2} < 400$ MHz, with a filter fixed in [0–400] MHz
7	$f_{s1} < 2$ GHz, $f_{s2} < 400$ MHz, with a filter fixed in [0–400] MHz
8	$f_{s1} < 2$ GHz, $f_{s2} < 400$ MHz, with a filter fixed in [0–200] MHz

Table 3. Subsampling receiver's architectures.

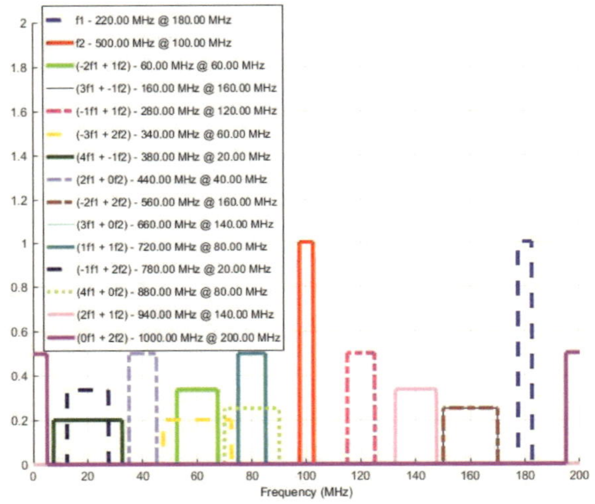

Figure 10. Simulated spectra after two-stage subsampling process.

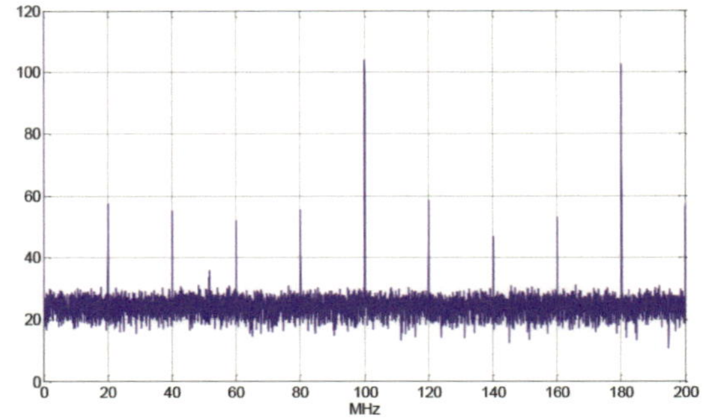

Figure 11. Experimental spectra after two-stage subsampling process.

captured output spectrum using the experimental setup. Observe that, after the first S&H, the RF signals have been down-converted to 220 and 500 MHz. After the second S&H, they are translated to 180 and 100 MHz, respectively. The subsampled signals, their harmonics, and intermodulation products are located as predicted by the simulation illustrated in **Figure 10**.

Finally, **Figure 12** shows the measured SNR for 12 different dual-band signal scenarios [15].

Using a multiplexed bank of filters (RD 5) presents more advantages and leads to better signal quality than receiver designs that use fixed filters (RD 6/7). The pros for these filter architectures (either fixed or multiplexed) are the elimination of a higher number of nonlinear components and the reduction of the out-of-band noise in the input ADC. However, when the carrier RF separation for the two bands is relatively small, it is difficult to find a large sampling frequency that folds both bands into one band-pass filter and, therefore, the effect of folded noise increases. It is possible to observe this problem in some dual-band scenarios using RD 8. Taking into account all these considerations, dual filter band-pass architecture seems to be an effective solution that offers reasonable performance without a sharp increase in the complexity (number of filters) for well-frequency-spaced dual-band wireless applications.

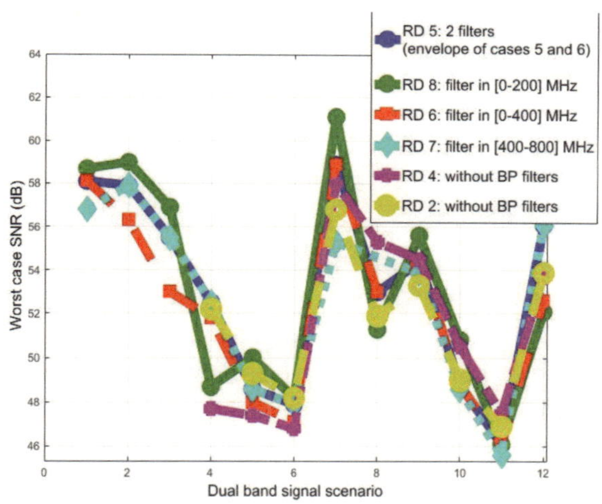

Figure 12. Experimental SNR for the proposed architectures.

2.4. Subsampling-based systems for cognitive radio applications (II): integration with compressive sensing techniques

2.4.1. Fundamentals of compressive sensing

Compressive sensing is an emerging alternative to Nyquist sampling for the acquisition of sparse signals [16]. The general idea is that the spectral information of a signal is not necessarily as large as its bandwidth occupation. This paradigm has been also named analog-to-information (AI) conversion [17], since the receiver is designed considering the mathematical structure of the signals rather than their bandwidth. The AI converter output is not the received waveform, but the minimum number of data to encode the information of interest, that is, its compressed version.

By using CS principles, it is possible to implement a reliable reconstruction of the signal of interest (SOI) from a reduced number of samples. Instead of taking periodic samples, CS measures products with M properly chosen measurement vectors, where M is much smaller than the number of Nyquist rate samples. Mathematically, let $X = [x_1, x_2,..., x_N]$ be the input signal, where x_i are its individual samples and N is the number of samples. The CS measurements y_k are the projections of X onto a measurement vector φ_k. Formally, $y_k = (X, \varphi_k)$ for $k = 1, 2, ..., M$, where M is the number of measurements [18]. For this technique to be useful, the signal has to be sparse in the basis $\{\varphi_k\}$, that is, the signal is expected to have a limited number of nonzero components when they are represented or projected in that basis. Although real-world signals are not totally sparse, as they cannot be completely band-limited and there is always noise embedded in the bandwidth, in many cases it is possible to obtain accurate approximations. In the end, CS enables us to reduce the requirements at the receiver and operate at a low data rate. However, CS also has two important challenges: (1) to select the appropriate number M of sub-Nyquist samples and measurement vectors, and (2) the need to carry out computationally expensive algorithms to recover the signal from its projections. Note, however, that we do not need to implement these algorithms with analog circuitry; rather, they can be performed in the digital domain.

2.4.2. *Optimization of compressive sensing architectures: integration with subsampling-based systems for scanning of wideband spectrum application*

Due to the underutilization of the radio spectrum, its occupancy can be usually considered sparse. Under this assumption, CS technique is a feasible alternative to many cognitive radio applications, where a reaction in real time is necessary, such as dynamic spectrum sensing [19, 20], interferer's mitigation [21], power spectrum estimation [22], or sparsity order estimation [23]. All these applications are focused on identifying spectrum opportunities over a wideband spectrum rapidly and accurately, so that we can share and exploit these limited radio resources in real time.

Although the computational burden is pretty high for CS techniques, there are many research works that aim to reduce this complexity, mainly at signal processing level. As an alternative, in this section, we address some ideas to reduce the complexity at circuit level, proposing the integration of compressive sensing with the subsampling algorithms proposed in the previous sections. The final objective is to integrate the advantages of both techniques, that is, a few number of samples for each SOI by using a cost-effective and flexible alternative receiver and avoiding the use of high-speed ADC circuitry.

Originally, the basic considered options to implement the CS receiver architectures for spectrum sensing applications have been homodyne and heterodyne receivers. A homodyne receiver with a wideband front end is able to implement rapid detections, but it implies the use of high-speed ADCs, which make these solutions impractical. A heterodyne receiver, where the signal channels are selected individually for downconversion to baseband, can be implemented by using a lower data-rate ADC, but they are impractical for spectrum sensing applications because the time to detect idle channels is higher than the homodyne case [19], due to the fact that heterodyne receivers consume time for switching to a new channel, apart from the time required by the detection algorithm, once the channel is selected. As a compromise solution, subsampling-based receivers enable us to use a low-cost ADC and provide a fast spectrum scanning [19].

Previously to the idea of using subsampling-based systems for CS applications, several receiver architectures have been proposed, which can be classified into two classes: nonuniform samplers and random (or pseudo-random) pre-integrators (**Figure 13**) [17, 24]. However, the performance of these sub-Nyquist schemes depends on the sparsity level, that is, the employed reconstruction algorithms will only work accurately if most of the channels in the spectrum are idle, leading in many cases to higher computational complexity and longer signal-processing times. Furthermore, although the ADCs in **Figure 13** work at a sub-Nyquist rate f_s, the detector requires analog components capable to work at Nyquist rate: for the nonuniform sampler, a Nyquist rate clock is required for the shifters (φ_i in **Figure 13a**); for the random pre-integrator, a Nyquist rate random generator ($p_i(t)$ in **Figure 13b** before the integrator $h(t)$) is required as well [19]. Besides the high-rate requirement for high bandwidth, a higher frequency of the digital parts in a mixed-signal system involves a higher injected noise into the analog parts. Consequently, subsampling-based systems emerge as a feasible alternative to implement CS receivers that avoid the operation at Nyquist rates.

In Ref. [19], a subsampling-based receiver is proposed to overcome the constraints detailed before, by sampling under Nyquist the input signal with a multiple branch architecture (**Figure 14a**), where there is a compromise between the number of branches and the scanning time. Although the results presented in Ref. [19] constitute a middle solution in terms of low-complex circuitry and fast detection, the proposed architecture improves the efficiency when searching idle slots. To this end, the methodology described in Section 2.3.3 for multi-band scenarios can be used with two main objectives:

1. Minimization of the number of branches in **Figure 14a**. As described in Section 2.3.5, applying the methods proposed in Ref. [15], it is possible to find the optimal band-pass filtering architecture, by searching the valid set of sampling frequencies in order to avoid aliasing with the interferers (**Figure 14b**), and also attending to requirements of folded noise reduction by using the principles of multiple clocking techniques (Section 2.2). Therefore, a similar scheme integrated in **Figure 14a** could minimize the number of branches, since the possibilities of aliasing in multi-band scenarios are reduced.

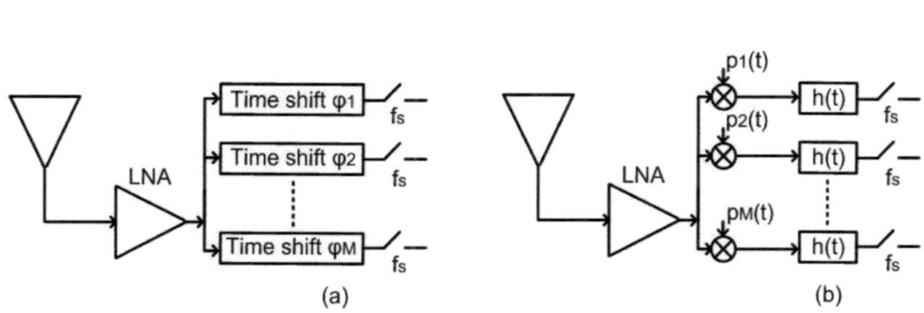

Figure 13. Nonuniform samplers (a) and random pre-integrator (b) architectures.

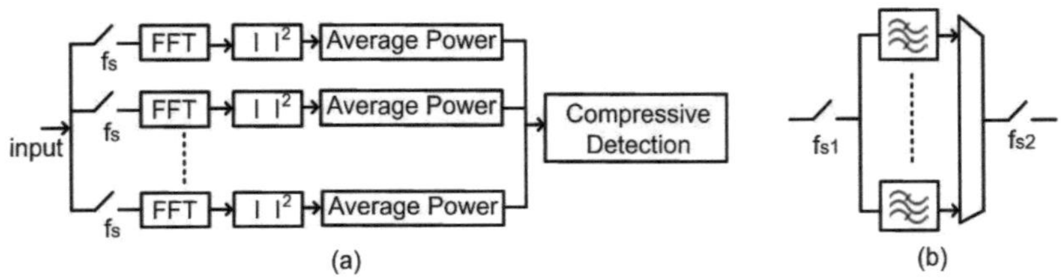

Figure 14. Subsampling architecture proposed in Ref. [19] (a) and Ref. [15] (b).

2. From the predefinition of these sets of valid frequencies, it would be possible to reduce the scanning times of idle slots, improving the efficiency of the architecture proposed in [19].

Finally, this approach is extensible to more realistic scenarios, by introducing the equations for nonlinear environments described in Section 2.3.4.

3. Interleaving techniques

3.1. Overview of the time-interleaved ADCs concepts

Time-interleaved ADCs (TI-ADC) are an effective approach for achieving very high sampling rates. A TI-ADC time-multiplexes M parallel ADCs with slightly delayed sampling instants, which is equivalent to a single ADC operating at a much higher sampling rate. The concept is illustrated in **Figure 15**. Ideally, the ith ADC, $i = 0,\ldots, M-1$, samples periodically the input signal at time instants t_i, t_{i+M}, t_{i+2M}, ... with sample rate f_s/M, where $t_m = mT_s$ and $T_s = 1/f_s$ is the sampling period of the TI-ADC. The final output is created by multiplexing all the individual ADC outputs in the proper order (e.g., ADC_0, ADC_1, ..., ADC_{M-1}, ADC_0, ADC_1, etc.). Thereby, the final effect is as if the input signals were sampled once every T_s seconds, that is, with sample rate f_s. This approach has been widely adopted in many communication systems for wideband spectrum applications, since the converters can work at lower speeds without sacrificing the overall system performance. However, it should be noted that each individual ADC deals with the entire analog input signal, and, therefore, its S&H circuit must be able to preserve the full input signal bandwidth.

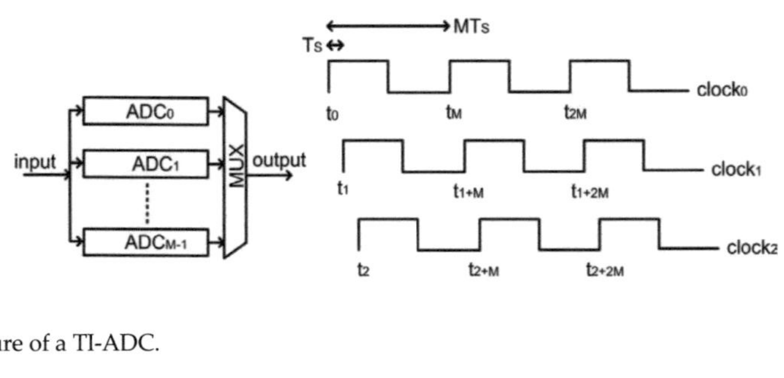

Figure 15. Architecture of a TI-ADC.

3.2. Analysis of time-interleaved ADCs: mismatch errors and calibration techniques

3.2.1. Mismatch errors in interleaving-based systems

Let $x(t)$ be an analog signal with Fourier transform $X_a(\omega)$ and consider that the TI-ADC produces the sequence:

$$x(t_0), x(t_1), x(t_2), \ldots, x(t_m), \ldots, x(t_M), x(t_{M+1}), \ldots \tag{18}$$

The discrete-time Fourier transform is defined by[1]

$$X(\omega) = \sum x(t_k)\exp(-j\omega t_k) \tag{19}$$

Ideally, the samples are spaced $T_s$seconds apart. Then, it can be shown that

$$X(\omega) = f_s \sum X_a(\omega - k2\pi f_s) \tag{20}$$

which is the well-known spectrum representation of a uniformly sampled signal. It results in a periodic spectrum with a period equal to the sampling rate [25]. In practice, however, there are deviations from the ideal behavior that are caused by the mismatches between the individual ADCs. There are three main possible sources of error in TI-ADCs: *clock timing errors*, *gain errors*, and *offset errors*. A brief review of all of them is included in the subsequent sections.

3.2.1.1. Clock timing errors

Clock timing errors occur when the digitization clocks of the individual ADCs are not appropriately synchronized. As a result, the input signal $x(t)$ is sampled in such a way that the sampling time instances are not necessarily uniformly spaced in time. Errors may be systematic (skew) or random (jitter). Taking this factor into account, Eq. (20) becomes [25]

$$X(\omega) = f_s \sum H_k(\omega) X_a(\omega - k2\pi f_s/M) \tag{21}$$

where

$$H_k(\omega) = \frac{1}{M} \sum \exp(-j[\omega - k(2\pi/MT_s)]\, r_m T_s)\exp(-jkm(2\pi/M)) \tag{22}$$

where $r_m = (t_m - mT_s)/T_s$ is a ratio that measures the timing errors (ideally, $t_m = mT_s$). It can be noticed that, in contrast with the ideal case, the spectrum is repeated in Eq. (21) every integer multiple of the frequency f_s/M (not f_s). In other words, spurious replica spectra (called *image spurs*) will appear at

[1]In the literature, it is a common notational practice to replace $\omega t_k$with a single variable $\omega' = \omega t_k$, called *normalized frequency*. Since ω represents ordinary frequency (radians per second), ω' is expressed in units of radians (per sample). Recall also that by sampling the discrete-time Fourier transform, we obtain the discrete Fourier transform (DFT).

$$f_{\text{imagspurs}} = \pm f_i + k\frac{f_s}{M}, \quad k = \pm 1, \pm 2, \pm 3, \ldots \tag{23}$$

These replicas hamper the interpretation of the spectrum of the input signal, as they may be confounded with true signal's frequency components. In addition, even if the image spurs are eliminated, we find that the spectrum of the signal reconstructed from the given samples is equal to $H_0(\omega)X_a(\omega)$. This is equivalent to passing the original input signal through a filter of transfer function $H_0(\omega)$, which introduces distortion and must be corrected by an equalizer. Assuming that the input signal is a pure tone of frequency f_i, and that f_i is a Gaussian variable with zero-mean and variance σ^2 for all m, *th*e signal-to-noise and distortion ratio (SNDR) is approximately found to be [25]

$$\text{SNDR} = 20\log\left(\frac{f_s}{2\pi f_i}\right) - 10\log\left(1 - \frac{1}{M}\right). \tag{24}$$

3.2.1.2. Gain errors

The gains of each ADC may be different, implying that, for example, even for a DC input each ADC may produce different output codes. The analysis is similar as for timing errors. Gain errors also produce image spurs at

$$f_{\text{imagspurs}} = \pm f_i + k\frac{f_s}{M}, \quad k = \pm 1, \pm 2, \pm 3, \ldots \tag{25}$$

where f_i is the input frequency. A formula for the SNDR, taking into account the contribution of gain and offset errors, will be given a few lines below.

3.2.1.3. Offset errors

The offsets of each ADC may be also different and then, as with the gain mismatch case, even a DC input may produce different outputs. Offset errors cause noise peaks (*offset spurs*) at

$$f_{\text{offspurs}} = k\frac{f_s}{M}, \quad k = 0,\ 1,\ 2,\ 3, \ldots \tag{26}$$

Assuming that the gain and offset errors are Gaussian distributed, with respective variances σ^2_g and f_i, and that the input signal is a sinusoid with amplitude A, the SNDR equals to [26]

$$\text{SNDR} = 10\log\left(\frac{A^2}{A^2\sigma_a^2 + 2\sigma_b^2}\right) \tag{27}$$

which is independent of the degree of interleaving M. The image and offset spurs are illustrated by the example shown in **Figure 16**. The input signal is a sinusoid at a frequency of 2.3 GHz that is sampled at f_s= 6 GHz, and the TI-ADC has three channels (i.e., three individual ADCs in parallel). In practice, clock timing errors are the most critical (gains and offsets can be

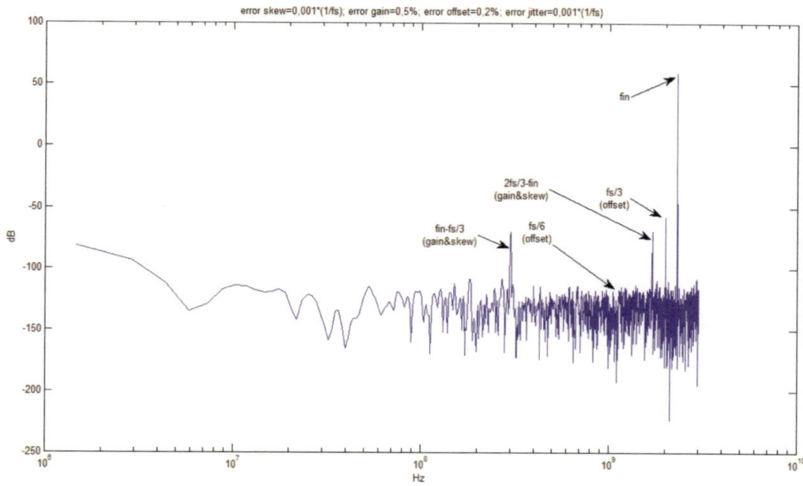

Figure 16. Output spectrum of a three-channel TI-ADC.

calibrated more easily). Several approaches have been proposed to deal with mismatches between individual ADCs. Some of the most relevant are presented in the subsequent section.

3.2.2. *Calibration techniques*

The mismatch errors described in the last section need to be corrected in order to maximize the benefits of interleaving systems and increase the ADC resolution, avoiding the degradation of the SNDR of the analog-to-digital system. To implement interleaving systems, several approaches, which differ in the calibration method and in the intended application, are possible. This section begins with a brief summary of the most prevalent calibration methods, discussing their advantages and disadvantages. Then, a review of several approaches for designing SDR, ultra wideband (UWB), and multistandard systems will be presented.

In Ref. [27], where an ADC system for UWB is described, two basic (nonexclusive) options to calibrate the system are presented: (1) controlling the ADC mismatches in the integrated circuit fabrication process and (2) using digital preprocessing or postprocessing techniques to mitigate the impact of image and offset spurs. The first one consists of reducing the electrical and physical differences between the channels. The gain is typically controlled using a common reference voltage and carefully designing the layouts. Phase matching is achieved by ensuring that all the clock paths are as similar in length as possible. Nevertheless, unlike approaches based on digital processing, these techniques are not well suited for implementations based on COTS (Commercial Off The Shelf), which is very convenient for SDR designs due to its flexibility and ease of programming.

Digital processing techniques have several advantages, including technology scaling, flexibility, the ease of portability to the next technologies generation, a low power consumption, and the possibility of being implemented through a cheap CMOS process. Moreover, digital techniques are more efficient for the compensation of timing skews than analog techniques, due to the fact that they are more stable with the temperature and the wide bandwidths used in SDR applications. Digital processing techniques are also easier to implement, can be designed with more precision, and take advantage of the last advances in high-speed and configurable digital hardware platforms (DSPs, FPGAs, CPLDs, and ASICs) [28].

In spite of the previously stated digital processing advantages, analog calibration techniques have benefits as well. For example, one advantage of analog offset calibration is that the correction values do not suffer from quantization [29]. Thus, the ADC offset can be corrected to less than one LSB without adding the extra bits needed when the offset correction is performed in the digital domain [30]. Furthermore, analog gain correction can be implemented by simply adjusting the reference voltage (so that using high-speed multi-bit digital multipliers to scale the ADC outputs becomes unnecessary).

Calibration techniques can be also classified into "background" and "foreground" techniques [31–33]. Sometimes, they are also named as, respectively, "online" and "offline" calibrations. Offline calibration requires less circuitry, but interrupts the normal operation of the ADC. It is usually applied when the parameters of the circuit do not vary much with environmental parameters (e.g., voltage or temperature). On the other hand, background techniques enable continuous calibration, with the ADC running in normal operation, and are suitable to be used when disconnecting the ADC is not an option [34].

A particularly appealing background approach is the one based on randomization [32]. In this approach, the ADC channel is selected randomly for each sampling instant. This can be performed using a digital circuit (thus avoiding the need for new analog circuitry), which constitutes the main benefit of this approach. Considering $M + X$ individual ADCs, we can choose at each time instant among $X + 1$ channels without violating the sampling rate of each individual ADC. This is illustrated in **Figure 17a** and **b**. In **Figure 17a** [35], a 3-ADCs system is presented with $M = 3$ and $X = 0$: noting that each ADC samples every 3 T_s seconds, where T_s is the sampling period of the total system, the only possible outputs are *A B C A B C A B C*… or *A C B A C B A C B*… For example, considering the first case, after the three channels (*A*, *B*, *C*) have taken a sample we can only choose *A* without violating our sampling constraint. Next, only *B* can be selected and so on. Therefore, to enable the randomization process, one or more extra ADCs (i.e., $X > 0$) must be employed [35]. In this way, there always exist at least two available ADCs at each sampling time. This case is illustrated in **Figure 17b**, which considers the case $M = 2$ and $X = 1$: after the first two channels have been selected, for example, *A* and *B*, we can decide between *A* and *C* and so forth.

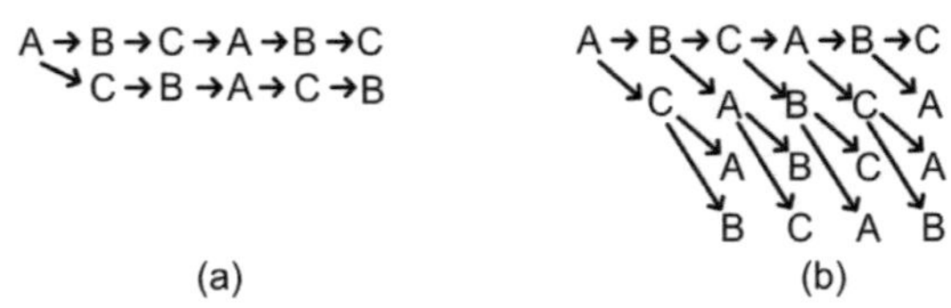

Figure 17. (a) 3-ADCs (*ABC*) for three times sampling rate and (b) 3-ADCs (*ABC*) for a double sampling rate.

An example of this architecture is illustrated in **Figure 18a**, where ΔM extra ADCs have been added.

Although this structure implies an additional hardware cost, Dyer et al. [35] demonstrates that as X tends to zero (in order to reduce the cost), noise becomes non-white (it is not flat). Therefore, a trade-off between the value of X and the cost of having nonactivated ADCs (note that it is not convenient to have largely underutilized ADCs, which occurs when X is much larger than M) is necessary. A clock diagram for $M = 4$ and $X = 1$ is shown in **Figure 18b** [35]. The sequence of randomly chosen ADCs is shown at the top of the figure.

Apart from the use in randomization techniques, an extra ADC might be employed in other background approaches. Doris et al. [33] use an additional ADC to implement an analog background circuit which calibrates the gain and offset mismatch. The basic idea of this technique is to use $M + 1$ ADCs so that M ADCs are always active while the remaining one is being calibrated. When the calibration cycle is finished, another ADC is selected for calibration, being replaced in the conversion mode by the previously calibrated ADC. The calibration of these ADCs is performed using another ADC as a reference, in order to match gain and offset with it. Therefore, an analog calibration is implemented. Since this method is based on a background process, it is not necessary to stop the analog-to-digital conversion and, in addition, it is more efficient as the number of ADCs increases. However, the main drawbacks are the noise introduced by the additional analog circuitry, and the degradation in speed when extra ADCs are used to substitute the ADC under calibration [36].

Moreover, there are also calibration techniques specifically devised either for static (i.e., gain and offset) or dynamic errors (i.e., clock skew). For example, an alternative method for calibrating the static mismatches is to employ a DAC (digital-to-analog converter) with enough resolution to meet the requirements [37], using one of these DACs for each ADC in the interleaved structure. A common alternative to reduce the cost of the extra circuitry due to the added DACs is to implement the gain and offset calibrations after a FFT (fast Fourier transform)

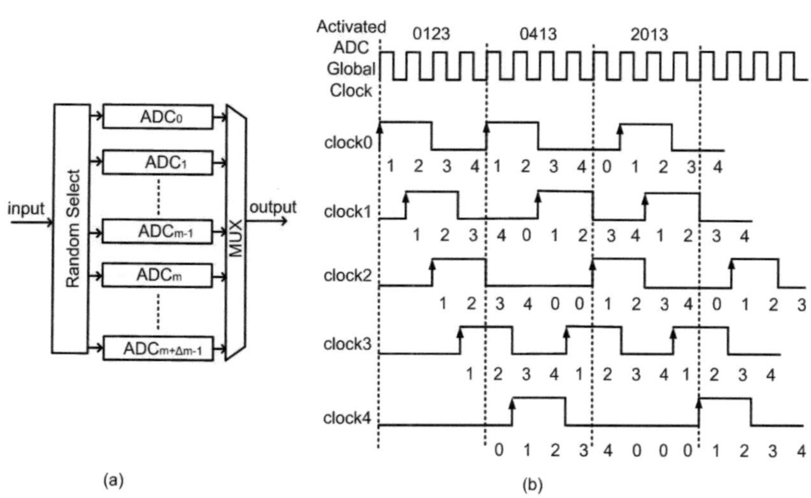

Figure 18. (a) Example of a structure with extra ADCs and (b) random clock for 5 ADCs.

evaluation, via software, achieving the compensation from the study of the output spectrum [38]. An additional benefit of this technique is to take the advantage of the repetition and symmetry properties of the FFT [39].

On the other hand, clock skew errors are difficult to correct and lead to more stringent limitations on interleaved architectures [40], especially for SDR and UWB applications as explained in Ref. [27]. Unlike static errors, dynamic errors depend on the input frequency and can be theoretically characterized from the aperture jitter and aperture delay of the internal S&H of each ADC. Although the aperture jitter affects high-frequency systems, its value (less than 1 ps) is usually much lower than the delay jitter (a few nanoseconds) and, therefore, it may be ignored in many interleaving-based systems. Moreover, the aperture delay strongly depends on the temperature. As a consequence, there are many unknown factors affecting these timing mismatches that make the estimation of the spurious component level even harder.

Some traditional ideas for correcting the skew errors are to add programmable delay lines or to implement a signal postprocessing. Dyer et al. [35] propose this method to reduce the spurious level though additional hardware is required. Another specific technique for timing errors correction is the one based on polyphase digital filter blocks [41], which can be easily implemented using an FPGA and, therefore, results highly suitable for SDR applications based on COTS. An example of this structure is illustrated in **Figure 19**, where the FPGA includes, as well as the filter implementation, a precision voltage reference, a low jitter clock distribution circuit, and a digital sensor to control the effect of temperature on the skew [28].

The filter structures used to calibrate these dynamic errors are designed as a function of the number of channels (i.e., number of ADCs). Reference [41] gives the ratio (R) between the number of channels (M) and the number of necessary filters (N), this ratio being lower when the number of channels is increased. For instance, in this work $N = 3$ for $M = 2$ ($R = 0.75$), $N = 5$ for $M = 3$ ($R = 0.556$), and $N = 27$ for $M = 8$ ($R = 0.422$). As a consequence, for a high number of ADCs this method is not efficient. Another inconvenience is that this technique is limited to a

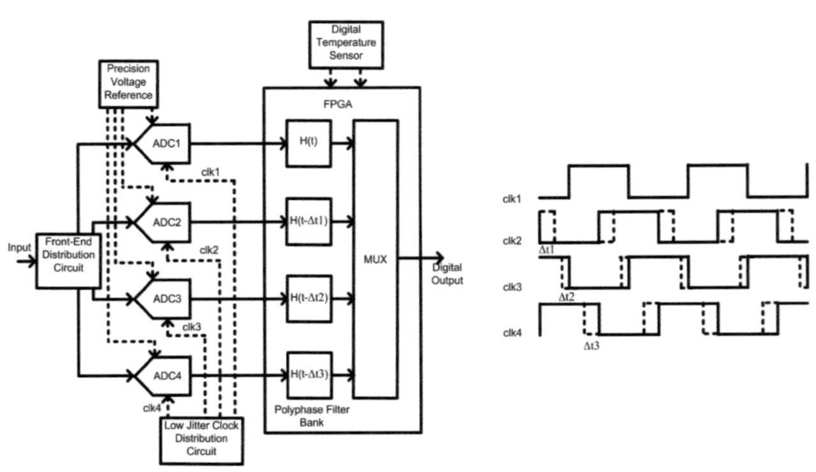

Figure 19. Functional diagram using digital filters blocks.

single-frequency input signal, and the filter coefficients would have to be frequently recalculated, thus increasing the calibration process complexity [42]. Lastly, an additional drawback of this method is that conventional filters do not have enough resolution to tune the group delay in the highest frequency systems, where the calibration precisely has to be more accurate [43]. Thus, designing these filters is an important challenge in their own.

3.3. Interleaving for cognitive radio applications: implemented systems and integration with subsampling and compressive sensing techniques

Placing the ADC interface as close to the antenna as possible in the signal path enables direct sampling of broadband multi-carrier signals of different standards with complex modulation schemes. Interleaving-based ADCs are a feasible alternative in order to achieve these objectives of cognitive radio, mainly for the base-station receivers, where the whole bandwidth has to be converted.

Many research works have been published in order to implement receivers and ADC systems based on interleaving techniques, most of them with application in SDR. Some of these systems are implemented in IC, whereas others are implemented with COTS, leading to some benefits for SDR applications, such as significant time and cost savings compared with developing an integrated circuit.

A very relevant time-interleaving system based on COTS is presented in Ref. [27]. In this system, eight ADCs MAX104 sampled at 1 GSPS are connected to obtain a total sampling frequency equal to 8 GHz, although, since the digital processing is implemented on a FPGA, the final frequency is equal to 6.4 GHz, due to speed restrictions of the device. Even though this is an inconvenience for high-speed circuits, the FPGA provides the necessary flexibility for SDR and UWB applications. Moreover, this solution means a reasonable tradeoff between cost and complexity. The proposed system presented in Ref. [27] employs a digital calibration scheme based on filters, obtaining an SNR around 30–35 dB (i.e., around 5–6 bits) at the maximum operation frequency.

Other systems based on interleaving techniques are not implemented by COTS but integrating all the system, including the calibration part (calibration on-chip). This solution is more convenient for analog or DAC-based calibration, reducing the power consumption [32, 44]. In [32], four ADCs are connected to obtain an SNDR around 48 dB for a total sampling frequency of 2.6 GHz. In Ref. [44], two ADCs are combined to achieve an SNDR higher than 54 dB for a sampling at 1 GHz.

There are other TI-ADCs which combine analog and digital calibration. For instance, Tamba et al. [34] use DACs to correct offset errors in the analog domain and background digital calibration to reduce the clock-skew effects. Using both methods, this work implements an 8-ADCs array to obtain a resolution of 5 bits at 12 GS/s operation rate. A similar work with cognitive radio applications is described in [45], where a resolution of 5.1 bits is obtained with 8 TI-ADCs operating at 10.3 GS/s.

Finally, there are some TI-ADCs that have applied successfully background digital calibration for dynamic and static errors compensation, as [46], which obtains 6 bits at 16 GS/s for eight different channels, and [47] that presents a wide-band analog-to-digital system for cognitive radio applications as well.

On the other hand, an important issue about the implementation of TI-ADCs is the connection between the S&Hs and ADCs. There are two possible structures, which are illustrated in **Figure 20** [40]. One of them uses only one S&H connected to several ADCs (**Figure 20a**). In this case, the main inconvenience is that the number of ADCs connected to the S&H is limited and, therefore, since this number of ADCs is directly proportional to the total sampling rate, the scalability will be limited as well. When each TI-ADC is preceded by its own S&H (**Figure 20b**), the scalability is theoretically increased because this architecture is not limited by a maximum number of ADCs connected to the S&H. However, the clock skew effect between the different S&Hs will degrade the final SNDR and, therefore, the final number of TI-ADCs will also be limited in this case, as it is shown in Eq. (24).

With the objective of reducing the number of S&Hs, Gupta et al. [40] propose a structure based on double sampling (**Figure 21**). Since using this solution the loading of the S&H is not directly dependent on the number of TI-ADCs, the scalability is increased with respect to the structure shown in **Figure 20a**. Besides, the mismatch errors and the power consumption are reduced with respect to the scheme shown in **Figure 20b**.

Additionally, interleaving techniques can be integrated with the previously detailed subsampling techniques with the objective of implementing SDR receivers that exploit the advantages of both technologies. For example, a high-resolution and high-analog bandwidth multiple clock subsampling-based system can improve its folded noise performance (Section 2.2) by using TI-ADCs to increase the second sampling rate f_{s2}. This solution would allow a higher intermediate frequency, which is folded by a higher first-sampling rate f_{s1}, reducing the thermal noise effect.

Integration of both techniques also can be used with other objectives. Louwsma et al. [48] propose a time-interleaved multichannel track and hold for a subsampling-based SDR receiver, with 48 dB of SNDR at 1.35 GS/s. This technique can be used to increase the total sampling rate as well. In this case, Louwsma et al. [48] propose to apply subsampling directly at the first track and hold. Furthermore, digital calibration techniques are usually employed only for band-limited signals in the first Nyquist band, so that they cannot be directly applied for band-pass-sampling schemes. Therefore, some research works are dedicated to calibration algorithms focused on undersampled signals. An example is Ref. [49], which validates the

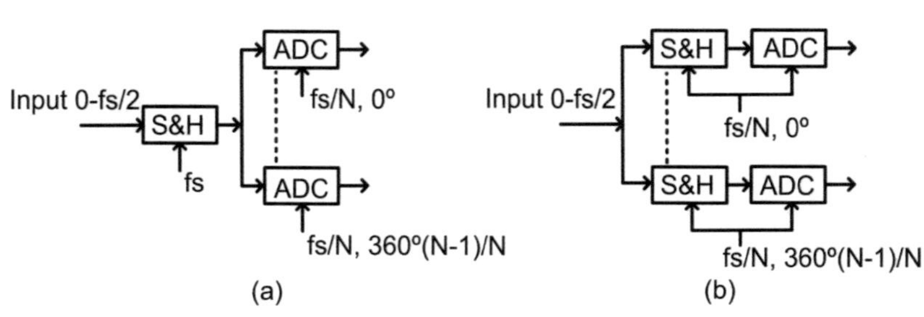

Figure 20. Architectures based on (a) one S&H and (b) several sub-S&H.

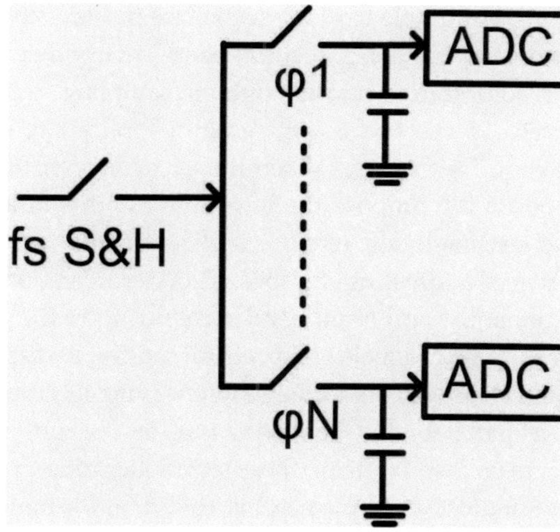

Figure 21. Architecture based on double sampling.

precision and convergence of its proposed digital background calibration method for clock-skew calibration of a two-channel subsampled TI-ADC. More recently, Duc et al. [50] has presented a novel digital background calibration for clock skew, by using a polyphase-filtering technique, which is applied to a four-subsampled channel TI-ADC. The system is clocked at 2.7 GHz, achieving an 11-bit resolution in a 5.3-GHz analog bandwidth, that is, the proposed calibration method is extended to the fourth Nyquist band. Since these receiver architectures are designed for SDR applications, the calibration efforts are usually focused on clock skew errors, since they increase with the input frequency and overshadow the effect of other mismatches for broadband inputs.

Finally, it is also remarkable that interleaving techniques have been recently incorporated to the compressive sensing systems described in Section 2.4.1, with the main objective of implementing fast spectrum sensing for cognitive radio applications. For example, Moon et al. [51] present some novel reconstruction algorithms for a 2-TI-ADC wideband sparse-based receiver, without requiring an accurate clock skew calibration and reducing the number of detection channels. However, other works are focused on calibration techniques, such as Ref. [52], where a clock skew adjustment method is proposed for sparse spectrum applications, or Ref. [53], which studied the influence of the mismatch errors, described in Section 3.2.1, in the quality of the reconstructed signal.

4. Conclusions

This chapter reviews several methods for implementing the analog-to-digital conversion stage to achieve the objectives of SDR and CR paradigms, becoming efficient alternatives to conventional receivers. In a first part, we describe how a subsampling-based system is an appealing option in order to implement tunable and cost-effective multistandard receivers for dynamic spectrum access applications, detailing a multiple clocking technique in order to reduce the folded noise effect. We extend this idea to nonlinear and multi-band environments for spectrum sensing applications, proposing an algorithm to find the optimal sampling frequency in order to avoid the nondesired interferences. As a second cognitive radio application, we introduce techniques based on compressive sensing for the cases when the occupancy of the spectrum is very low, meeting the sparsity property. We propose the integration with subsampling architectures to implement a compressive sensing-based receiver and reduce the complexity of the receiver. Therefore, with the objective of addressing the spectral underutilization problem, we apply the algorithms proposed for nonlinear and multi-band scenarios to reduce the computational burden and the searching time of idle channels for spectrum sensing and spectrum sharing applications. The second part of the chapter is dedicated to interleaving techniques, which are based on several ADCs connected in parallel with the objective of maximizing the sampling frequency. Although the resolution is decreased by using these techniques, due to the generated mismatch errors, their effects can be minimized implementing calibration techniques such as those proposed in this chapter. Finally, we describe the benefits of integrating interleaving-based systems with architectures based on subsampling and compressive sensing, with the objective to combine the advantages of these techniques in a unique analog-to-digital system.

Acknowledgements

This work was supported by the Spanish Ministry of Economy and Competitiveness, in part under project TEC2013-47286-C3-1-R, and in part under project TEC2016-80396-C2-2-R.

Author details

José Ramón García Oya[1]*, Fernando Muñoz Chavero[1] and Rubén Martín Clemente[2]

*Address all correspondence to: jose.garciaoya@gie.esi.us.es

1 Department of Electronics Engineering, University of Seville, Seville, Spain

2 Department of Signal Theory and Communications, University of Seville, Seville, Spain

References

[1] Mitola J. The software radio architecture. IEEE Communications Magazine. 1995;**33**(5):26–38. DOI: 10.1109/98.788210

[2] Haykin S. Cognitive radio: Brain empowered wireless communications. IEEE Journal on Selected Areas in Communication. 2005;**23**(2):201–220. DOI: 10.1109/JSAC.2004.839380

[3] Mitola J, MaGuire GQ. Cognitive radio: Making software radios more personal. IEEE Personal Communications. 1999;**6**(4):13–18. DOI: 10.1109/98.788210

[4] Grace D, Pitt SP. Quadrature sampling of high frequency waveforms. Journal of the Acoustical Society of America. 1968;**44**:1432–1436. Doi: http://dx.doi.org/10.1121/1.1911284

[5] Vaughan R, Scott N, White D. The theory of bandpass sampling. IEEE Transactions on Signal Processing. 1991;**39**(9):1973–1984. DOI: 10.1109/78.134430

[6] Sun YR. Generalized bandpass sampling receivers for software defined radio [thesis]. Stockholm: School of Information and Communication Technology (ICT); 2006

[7] Karvonen S. Charge-domain sampling of high frequency signals with embedded filtering [thesis]. Finland: Faculty of Technology, Department of Electrical and Information Engineering, University of Oulu; 2006

[8] Oya JRG et al. Data acquisition system based on subsampling for testing wideband multistandard receivers. IEEE Transactions on Instrumentation and Measurement. 2011;**60**(9):3234–3237. DOI: 10.1109/TIM.2011.2128710

[9] Oya JRG et al. Data acquisition system base on subsampling using multiple clocking techniques. IEEE Transactions on Instrumentation and Measurement. 2012;**61**(8):2333–2335. DOI: 10.1109/TIM.2012.2200819

[10] Tseng CH. A universal formula for the complete bandpass sampling requirements of nonlinear systems. IEEE Transactions on Signal Processing. 2009;**57**(10):3869–3878. DOI: 10.1109/TSP.2009.2023356

[11] Tseng CH, Chou SC. Direct downconversion of multiband RF signals using bandpass sampling. IEEE Transactions on Wireless Communications. 2006;**5**(1):72–76. DOI: 10.1109/TWC.2006.1576530

[12] Bassam SA, et al. Subsampling feedback loop applicable to concurrent dual-band linearization architecture. IEEE Transactions on Microwave Theory and Techniques. 2012;**60**(6), part 2:1990–1999. DOI: 10.1109/TMTT.2012.2192745

[13] Kwan A, Bassam SA, Ghannouchi FM. Sub-sampling technique for spectrum sensing in cognitive radio. In: IEEE Radio and Wireless Symposium (RWS'2012); January 2012; Santa Clara, California: IEEE; 2012. pp. 347–350. DOI: 10.1109/RWS.2012.6175350

[14] Oya JRG et al. Optimization of subsampling dual band receivers design in nonlinear systems. In: IEEE MTT-S International Microwave Symposium Digest (IMS'2012); June 2012; Montreal, Canada: IEEE; 2012. pp. 1–3. DOI: 10.1109/MWSYM.2012.6259729

[15] Oya JRG et al. Design of dual-band multistandard subsampling receivers for optimal SNDR in nonlinear and interfering environments. IEEE Transactions on Instrumentation and Measurement. 2014;**63**(4):981–983. DOI: 10.1109/TIM.2013.2297651

[16] Candes E. Compressive sampling. In: Proceedings of the International Congress of Mathematics; August 2006; Madrid, Spain; 2006. pp. 1433–1452. DOI: 10.4171/022-3/69

[17] Laska J et al. Theory and implementation of an analog-to-information converter using random demodulation. In: IEEE International Symposium on Circuits and Systems; May 2007; New Orleans, LA, USA: IEEE; 2007. pp. 1959–1962. DOI: 10.1109/ISCAS.2007.378360

[18] Donoho DL. Compressed sensing. IEEE Transactions on Information Theory. 2006;**6** (4):1289–1306. DOI: 10.1109/TIT.2006.871582

[19] Bai L, Roy S. Compressive spectrum sensing using a bandpass sampling architecture. IEEE Journal on Emerging and Selected Topics in Circuits and Systems. 2012;**2**(3):433–442. DOI: 10.1109/JETCAS.2012.2214874

[20] Jiang J et al. Achieving autonomous compressive spectrum sensing for cognitive radios. IEEE Transactions on Vehicular Technology. 2016;**65**(3):1281–1291. DOI: 10.1109/TVT.2015.2408258

[21] Lagunas E, Najar M. Spectral feature detection with sub-Nyquist sampling for wideband spectrum sensing. IEEE Transactions on Wireless Communications. 2015;**14**(7):3978–3990. DOI: 10.1109/TWC.2015.2415774

[22] Ariananda DD, Leus G. Compressive wideband power spectrum estimation. IEEE Transactions on Signal Processing. 2012;**60**(9):4775–4789. DOI: 10.1109/TSP.2012.2201153

[23] Sharma SK, Chatzinotas S, Ottersten B. Compressive sparsity order estimation for wideband cognitive radio receiver. IEEE Transactions on Signal Processing. 2014;**62**(19):4984–4996. DOI: 10.1109/ICC.2014.6883511

[24] Chen X, et al. A sub-Nyquist rate sampling receiver exploiting compressive sensing. IEEE Transactions on Circuits and Systems I: Regular Papers. 2011;**58**(3):507–520. DOI: 10.1109/TCSI.2010.2072430

[25] Jenq YC. Digital spectra of nonuniformly sampled signals: Fundamentals and high-speed waveform digitizers. IEEE Transactions on Instrumentation and Measurement. 1988;**37** (2):245–251. DOI: 10.1109/19.6060

[26] Petraglia A, Mitra SK. Analysis of mismatch effects among A/D converters in a time-interleaved waveform digitizer. IEEE Transactions on Instrumentation and Measurement. 1991;**40**(5):831–835. DOI: 10.1109/19.106306

[27] Anderson CR, et al. Analysis and implementation of a time-interleaved ADC array for a software-defined UWB receiver. IEEE Transactions on Vehicular Technology. 2009;**58** (8):4046–4063. DOI: 10.1109/TVT.2009.2021268

[28] Looney M. Advanced digital postprocessing techniques enhance performance in time-interleaved ADC systems. Analog Dialogue. 2003;**37**(3):5–9

[29] Haykin S. Adaptive Filter Theory. Upper Saddle River, NJ: Prentice-Hall; 1986. ISBN:0-13-004052–5

[30] Fu D et al. Digital background calibration of a 10-b 40-MS/s parallel pipelined ADC. In: Proceedings of the IEEE ISSCC; San Francisco, CA: IEEE; 1998. pp. 140–142. DOI: 10.1109/ISSCC.1998.672407

[31] El-Sankary K, Sawan M. High resolution self-calibrated ADCs for software defined radios. In: The 16th International Conference on Microelectronics (ICM 2004); December 2004; Tunis, Tunisia; 2004. pp. 120–123. DOI: 10.1109/ICM.2004.1434223

[32] Doris K et al. A 480 mW 2.6 GS/s 10b time-interleaved ADC with 48.5 dB SNDR up to Nyquist in 65 nm CMOS. IEEE Journal of Solid-State Circuits. 2011;**46**(12):2821–2833. DOI: 10.1109/JSSC.2011.2164961

[33] Dyer KC et al. An analog background calibration technique for time-interleaved analog-to-digital converters. IEEE Journal of Solid-State Circuits. 1998;**33**(12):1912–1919. DOI: 10.1109/4.735531

[34] El-Chammas M, Murmann B. A 12-GS/s 81-mW 5-bit time interleaved flash ADC with background calibration. IEEE Journal of Solid-State Circuits. 2011;46(4):838–847. DOI: 10.1109/JSSC.2011.2108125

[35] Tamba M, et al. A method to improve SFDR with random interleaved sampling method. In: International Test Conference 2001; Baltimore, MD, USA: IEEE; 2001. pp. 512–520. DOI: 10.1109/TEST.2001.966669

[36] Ingino J et al. A continuously calibrated 12-b, 10-MS/s, 3.3-V A/D converter. IEEE Journal of Solid-State Circuits. 1998;**33**(12):1920–1931. DOI: 10.1109/4.735532

[37] Brown J, Hurst P, Der L. A 35 Mb/s mixed-signal decision feedback equalizer for disk drives in 2-μm CMOS. IEEE Journal of Solid-State Circuits. 1996;**31**(9):1258–1266. DOI: 10.1109/4.535409

[38] Pereira JMD et al. An FFT-based method to evaluate and compensate gain and offset errors of interleaved ADC systems. IEEE Transactions on Instrumentation and Measurement. 2004;**53**(2):423–430. DOI: 10.1109/TIM.2004.823321

[39] Slim HH, Russer P. Digital automatic calibration method for a time-interleaved ADCs system used in time-domain EMI measurement receiver. In: IEEE International Symposium on Electromagnetic Compatibility. (EMC 2011); August 2011; Long Beach, CA: IEEE; 2011. pp. 476–479. DOI: 10.1109/ISEMC.2011.6038358

[40] Gupta SK, Inerfield MA, Wang J. A 1-GS/s 11-bit ADC With 55-dB SNDR, 250-mW power realized by a high bandwidth scalable time-interleaved architecture. IEEE Journal of Solid-State Circuits. 2006;**41**(12):2650–2657. DOI: 10.1109/JSSC.2006.884331

[41] Lee YS, An Q. Design the efficient block digital filters for calibration of timing-error effects in time-interleaved ADC system. In: International Conference on Communications, Circuits and Systems 2005; May 2005; Hong Kong, China: IEEE; 2005. p. 2. DOI: 10.1109/ICCCAS.2005.1495360

[42] Abbaszadeh A, Dabbagh-Sadeghipour K. An efficient postprocessor architecture for channel mismatch correction of time interleaved ADCs. In: Proceedings of ICEE; May 2010; Isfahan, Iran: IEEE; 2010. pp. 382–385. DOI: 10.1109/IRANIANCEE.2010.5507040

[43] Asami K et al. Timing skew compensation technique using digital filter with novel linear phase condition. In: IEEE International Test Conference (ITC); November 2010; Austin, TX: IEEE; 2010; pp. 1–9. DOI: 10.1109/TEST.2010.5699234

[44] Lin Y, et al. An 11b 1GS/s ADC with parallel sampling architecture to enhance SNDR for multi-carrier signals. In: Proceedings of the Solid-State Circuits European Conference (ESSCIRC 2013); September 2013; Bucharest, Romania: IEEE; 2013. pp. 121–124. DOI: 10.1109/ESSCIRC.2013.6649087

[45] Nazemi A, et al. A 10.3GS/s (5.1 ENOB at Nyquist) time-interleaved/pipelined ADC using open-loop amplifiers and digital calibration in 90nm CMOS. In: 2008 Symposium on VLSI Circuits Digest of Technical Papers; June 2008; Honolulu, Hawaii: IEEE; 2008. pp. 19–20. DOI: 10.1109/VLSIC.2008.4585935

[46] Huang CC, Wang CY, Wu JT. A CMOS 6-bit 16-GS/s time-interleaved ADC using digital background calibration techniques. IEEE Journal of Solid-State Circuits. 2011;**46**(4):848–858. DOI: 10.1109/JSSC.2011.2109511

[47] Papari B, Asemani D, Khakpour A. A wide-band time-interleaved A/D converter for cognitive radio application with adaptive offset correction. In: 2011 Wireless Advanced; June 2011; London, UK: IEEE; 2011. pp. 144–148. DOI: 10.1109/WiAd.2011.5983302

[48] Louwsma SM, et al. A time-interleaved track & hold in 0.13μm CMOS sub-sampling a 4 GHz signal with 43dB SNDR. In: IEEE Custom Integrated Circuits Conference; September 2007; San Jose, CA: IEEE; 2007. pp. 329–332. DOI: 10.1109/CICC.2007.4405745

[49] Centurelli F, Monsurro P, Trifiletti A. Efficient digital background calibration of time-interleaved pipeline analog-to-digital converters. IEEE Transactions on Circuits and Systems I: Regular Papers. 2012;**59**(7):1373–1383. DOI: 10.1109/TCSI.2011.2177003

[50] Duc HL, et al. Fully digital feedforward background calibration of clock skews for sub-sampling TIADCs using the polyphase decomposition. IEEE Transactions on Circuits and Systems I: Regular Papers. 2017; PP(99):1–14. DOI: 10.1109/TCSI.2016.2645978, in press.

[51] Moon T, et al. Wideband sparse signal acquisition with dual-rate time-interleaved undersampling hardware and multicoset signal reconstruction algorithms. IEEE Transactions on Signal Process. 2015;**63**(24):6486–6497. DOI: 10.1109/TSP.2015.2469648

[52] Liu SJ, Xu XJ, Zou YX. Blind timing skew estimation based on spectra sparsity and all phase FFT for time-interleaved ADCs. In: IEEE International Conference on Digital Signal Processing; July 2015; Singapore: IEEE: 2015. pp. 926–930. DOI: 10.1109/ICDSP.2015.7252012

[53] Rial RM, Carrera JMC, Prelcic NG. Sensitivity of compressive sensing architectures based on time interleaved analog-to-digital converters to channel mismatches. In: IEEE 12th International New Circuits and Systems Conference (NEWCAS); June 2014; Trois-Rivieres, Canada: IEEE; 2014; pp. 45–48. DOI: 10.1109/NEWCAS.2014.6933981

Chapter 5

Reconfigurable Antennas for UWB Cognitive Radio Communication Applications

Yingsong Li and Yanyan Wang

Additional information is available at the end of the chapter

http://dx.doi.org/10.5772/intechopen.69079

Abstract

In this chapter, reconfigurable antennas are reviewed for ultra-wideband (UWB) cognitive radio communication applications. The defected microstrip structure (DMS) has been reviewed and integrated into the UWB antennas to form the desired filtering antennas which can filter out unexpected narrowband signal interferences. Then, switches are incorporated into the filtering UWB antennas to construct the cognitive radio UWB (CR-UWB) antenna to make the antenna switch between the UWB antenna and band-notched UWB antenna. In these CR-UWB antennas, the DMSs are to give the desired notches while the switches are used for realizing the switchable characteristics. Several reconfigurable antennas and CR-UWB antennas are created and investigated. The results show that the designed CR-UWB antenna can switch between different modes, making it amazing for UWB, band-notched UWB, and multiband communication system applications.

Keywords: reconfigurable UWB antenna, cognitive radio, switchable antennas, frequency rejection antennas, filtering antennas

1. Introduction

With the development of the modern wireless communications, the demand for high data transmission rate and wide bandwidth has attracted much more attention in both academic and industrial fields [1–4]. Specially, a ultra-wideband (UWB) communication system covering a wide bandwidth, which ranges from 3.1 to 10.6 GHz, has been released by Federal Communications Commission (FCC) in 2002 for commercial UWB communications [5]. In sequel, a great number of UWB studies have been presented to build a practical communication system since the UWB system can provide high data rate and good resistance for multipath and jamming [6, 7]. To transmit and receive wireless signal, antennas are important and should be integrated into a wireless communication device. Then, UWB antenna designs have been widely investigated because a UWB antenna is one of the important components to create a UWB communication system.

Many UWB antennas have been put forward to carry out wide bandwidth in the past few years, which includes uniplanar and planar antennas [8–12]. In sequel, planar microstrip antennas have been extensively developed because of their low cost and easy fabrication. Then, a lot of UWB antennas have been presented with different structures and various feedings [11–13], such as coaxial feed, microstrip feed, and coplanar waveguide (CPW) feeds. Many UWB antennas are realized by using microstrip feeds which can provide a wide bandwidth. However, some of these antennas are complex in structure and others are still embarrassed in narrow bandwidth. Then, many bandwidth enhancement techniques have been reported to expand the bandwidth of the existing antennas [14, 15]. Furthermore, UWB antennas with CPW feedings have been widely studied to achieve wide bandwidths and good omnidirectional radiation patterns. Although these UWB antennas can well cover the entire UWB bandwidth, they may be interfered by the existing narrowband wireless communication systems such as wireless local area network (WLAN) and worldwide interoperability for microwave access (WiMAX) [16–19]. After that a great number of UWB antennas with band notched functions have been reported to give resistance to these potential interferences by using various notch techniques [16–19], including U-slot, C-slot, L-slot, and E-shaped slots. These etched slots on the ground plane or radiation patches might leaky electromagnetic waves which may affect the electromagnetic compatibility designs of wireless communication systems [16]. Then, some improved band-notched UWB antennas with stubs and resonators have been presented to overcome the drawbacks which are caused by the leaky electromagnetic waves [12, 20–24]. These band-notched UWB antennas can well filter out the unwanted narrowband interferences. However, it cannot provide a good service when an entire banded UWB antenna is desired for sensing demands. Thus, a multimode antenna is desired to fulfill the mentioned communication requirements. A reconfigurable antenna is a good candidate for providing multiple modes. In fact, the reconfigurable antennas have been widely studied and used in various communications [25–32], such as cognitive radio systems. Moreover, the reconfigurable techniques can also be used for designing a cognitive radio antenna by implementing the UWB operating mode, band-notched UWB mode, and multiple band mode [25–32].

Recently, the reconfigurable UWB antennas have been developed for cognitive radio (CR) communications [25–32]. In the CR communication systems, unlicensed users (secondary users) can access spectrum bands licensed to primary users at a spectrum underlay mode or spectrum overlay mode, which is illustrated in **Figure 1** [28]. In the underlay mode, the secondary users are limited under a very low transmission power which is less than 41.3 dBm/MHz for UWB users [26]. This approach can be realized by using impulse radio (IR) based UWB (IR-UWB) technology [26]. For the overlay mode, the secondary users detect the existing narrowband (NB) signals, such as those signals from WLAN and radio frequency identification (RFID), and provide immunity to the NB systems. This can be implemented by turning off corresponding subcarriers in orthogonal frequency division multiplexing UWB (OFDM-UWB), depending on whether any primary users exist or not in a special band [26–28].

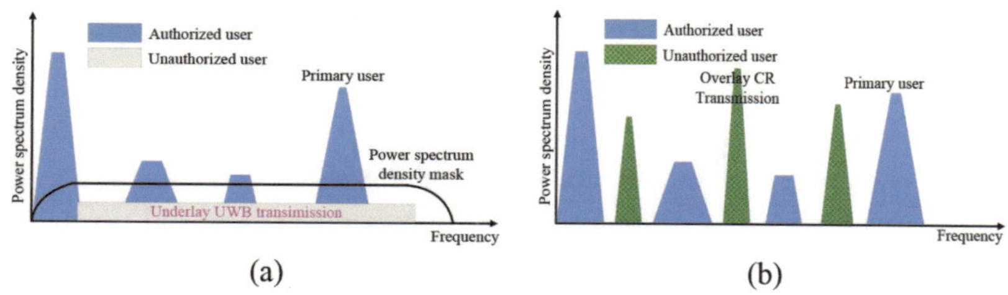

Figure 1. CR-UWB spectrum sharing modes. (a) Underlay. (b) Overlay.

In other words, the transmission spectrum of UWB radios can be sculpted in accordance with the presence of the primary users in the respective frequency bands [25–32]. Therefore, in CR-UWB systems, a CR antenna should cover the entire UWB band from 3.1 to 10.6 GHz with no notch bands for underlay applications and for detecting the licensed primary users and providing immunity to these users using band-notched technologies [25–32].

This chapter reviews the recent development of the CR-UWB antenna designs, including the defected microstrip structure (DMS)-based stop-band filter, band-notched UWB antenna and reconfigurable UWB antenna. First, DMS stop-band filter is briefly reviewed to give a discussion of the filtering UWB antenna designs. Second, the band-notched UWB antennas with stop-band filters are presented. Third, reconfigurable UWB antennas with multimodes are discussed to give an explanation to illustrate multiband and CR antenna designs. Finally, an example for developing CR-UWB antenna with wide bandwidth and multimode is introduced for multiband and CR communications.

2. Advances of UWB antenna designs

To provide a good service for CR-UWB communication system, a CR-UWB antenna is required to transmit and receive the desired signals [25–32]. The UWB communication system overlaps with the existing narrowband communication systems, such as WLAN, WiMAX, C-band, radio frequency identification (RFID), and X-band [16–19]. These communication systems occupy almost all the spectrum resources under the 10 GHz. Sometimes, we need to change the operation modes to meet multimode wireless communication and cognitive radio applications [25–32]. Thus, UWB antenna requires a signal selection scheme that can detect the presence of narrowband interferences [16–19, 25–32]. Then, the UWB antenna can provide effective notches to reject these potential interferences. One of the effective methods is to utilize CR-UWB technique. To construct a CR-UWB communication system, a CR-UWB antenna is desired to transmit and receive electromagnetic signals. Then, several CR-UWB antennas have been reported to meet the CR-UWB communication requirements [25–32]. However, the proposed CR antennas (CRAs) are complex in structure for dual port antennas [27, 29]. For the single UWB CRAs, the previous CRAs cannot be designed flexibly. Most of proposed single port cognitive antennas are designed using split-ring resonators (SRRs) in the radiation patch [26], which might leak electromagnetic wave that deteriorates the radiation patterns. Next, we will introduce a CR-UWB antenna design based on the band-stop filter techniques step by step.

2.1. Defected microstrip structure (DMS) band-stop filter

The DMSs have small size and can effectively give resistance to electromagnetic interferences (EMIs), which are carried out by etching various slots in the microstrip lines [33]. The DMSs can be used to filter out unwanted electromagnetic signals in a special frequency. Furthermore, the DMS is easy to fabricate and integrate into a microwave system and it can be effectively analyzed by using circuit theory [34]. A typical DMS, which is realized by etching a meander line slot on a 50-Ohm microstrip line, is shown in **Figure 2(a)** and its equivalent circuit model is given in **Figure 2(b)**. The performance of the DMS band-stop filter is shown in **Figure 3**. It is found that a stop band near 3.5 GHz is achieved, which can be used for filtering out the unexpected narrowband signals from the WiMAX communication system [34].

(a) (b)

Figure 2. A typical DMS band-stop filter and its equivalent circuit. (a) DMS band-stop filter. (b) Equivalent circuits.

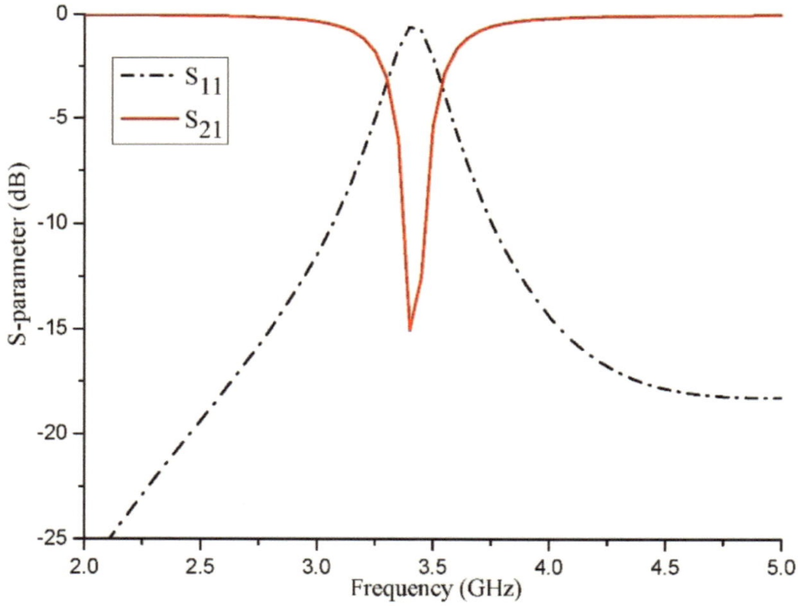

Figure 3. Performance of the DMS band-stop filter.

Then, another meander line slot is etched on the DMS filter given in **Figure 1** to create the second stop band. The configuration of the stop-band filter with two stop bands is illustrated in **Figure 4**. Two meander line slots with different dimensions are used to generate the desired stop bands [34], respectively. The larger meander line slot is to produce the lower stop band, while the smaller meander line slot can provide the upper stop band. The performance of the filter with two stop band is described in **Figure 5**. Another stop band is obtained around 5.25 GHz to reject the unwanted signal from the WLAN band. The two stop bands are given by using different DMS cells and they are controlled independently. Thus, we can embed different DMSs into the 50-Ohm microstrip line to construct multiple stop bands.

From the above discussions, it is observed that the DMS can be used for designing various stop bands. Thus, DMS-based stop-band filter can be used in a wireless system to reject the narrow-band signals. However, it will increase the complexity and the size of the wireless device since

Figure 4. Configuration of the filter with two stop bands.

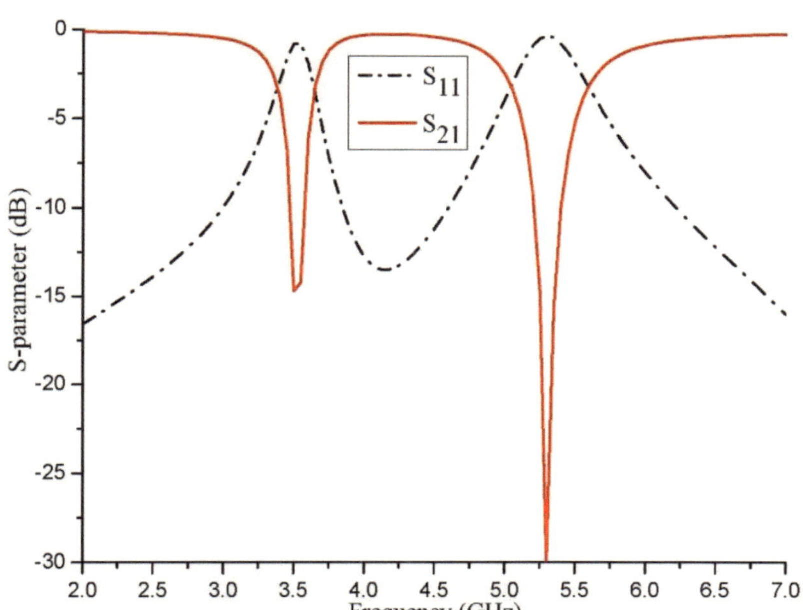

Figure 5. Behavior of the stop-band filter with two meander line slots.

it is usually added at the end of the antenna. Thus, band-notched UWB antenna can be realized by combining the stop-band filter and UWB antenna together to build a filtering UWB antenna.

2.2. Filtering UWB antenna

Here, the DMS stop-band filters are integrated into the microstrip feeding signal strip line to form filtering UWB antenna [34]. An example of the UWB antenna is shown in **Figure 6(a)**. The UWB antenna, which is printed on a substrate with a relative permittivity of 2.65, constants of a tapper patch, a ground plane, and a microstrip feeding signal strip line. The antenna can cover the entire UWB band ranging from 3.1 to 10.6 GHz with a voltage standing wave ratio (VSWR) <2 or a reflection coefficient <−10 dB. To prevent the potential interference from the WiMAX communication system at 3.5 GHz, a DMS cell is incorporated into the microstrip feeding signal line to create the filtering UWB antenna given in **Figure 6(b)**. In comparison with the UWB antenna shown in **Figure 6(a)**, there is an extra DMS cell in the feed transmission line.

The impedance characteristics of the UWB antenna and the filtering UWB antenna are demonstrated in **Figure 7**. **Figure 7(a)** shows the reflection coefficient (S11) and VSWR of the UWB

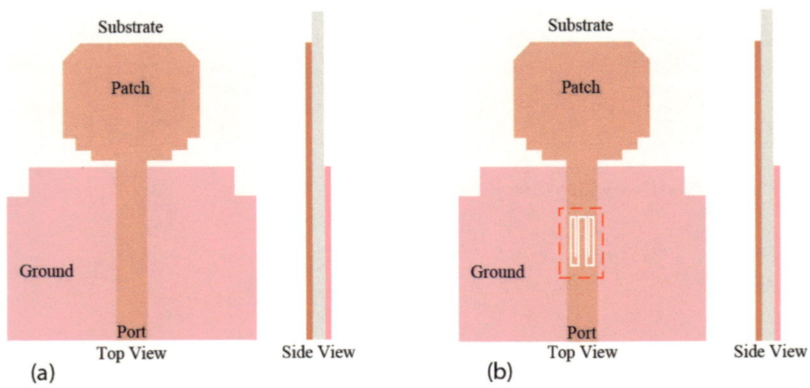

Figure 6. UWB antenna and filtering UWB antenna. (a) UWB antenna. (b) Filtering UWB antenna.

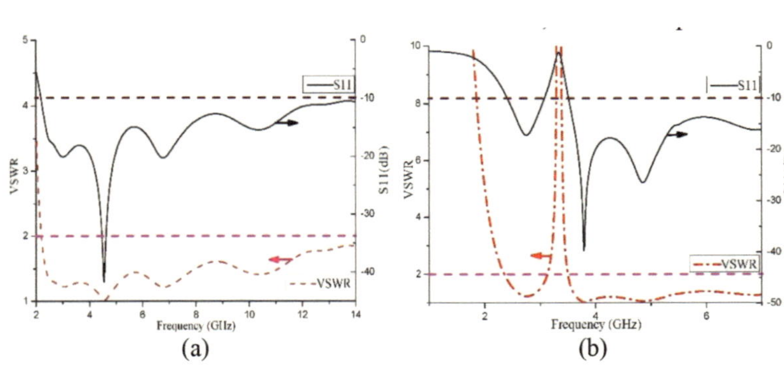

Figure 7. Impedance bandwidth of the UWB antenna and filtering UWB antenna. (a) Impedance of the UWB antenna. (b) Impedance of the filtering UWB antenna.

antenna, while **Figure 7(b)** presents the S11 and VSWR of the filtering UWB antenna. It can be seen from **Figure 7(a)** that the UWB antenna has a bandwidth of 147.5%, which covers the entire UWB band released by FCC. By incorporating a meander line slot into the transmission signal line of the UWB antenna, a notch is produced within the UWB band. The notch can well filter out the interferences from the 3.5 GHz WiMAX band. Also, the notch depth and bandwidth can be adjusted by changing the dimensions of the meander line slot. In this case, the stop-band filter and the UWB antenna are integrated together, which may reduce the size of the devices.

Then, another meander line slot is used for generating the second notch to filter out another narrowband interferences [26]. To better understand the independence of the notch bands, a UWB antenna with two meander line slots is created and its configuration is given in **Figure 8(a)**. These meander line slots are sequentially integrated into the feeding signal line. The VSWR and S11 are illustrated in **Figure 8(b)**. It is observed that a notch at 5.2 GHz is obtained by using the second meander line slot. Therefore, the two notch bands can be controlled by the meander line slots. As we know, these UWB antennas can well filter out the unwanted narrowband signal interferences. However, a new UWB antenna should be designed if the narrowband communication is necessary. Thus, a reconfigurable UWB antenna might be useful to switch between the UWB communication and band-notched UWB communication systems.

2.3. Reconfigurable UWB antenna

Since the CR-UWB antenna may change the modes between the band-notched UWB and UWB communication systems, a reconfigurable UWB antenna with a bandwidth of 2.38–7 GHz is presented to provide the desired switchable characteristics. Three switches, namely, switch-1 (SW1), switch-2 (SW2) and switch-3 (SW3), are incorporated into the band-notched UWB antenna to make the antenna reconfigurable. The reconfigurable antenna is shown in **Figure 9(a)** and its performance is given in **Figure 9(b)**. It is found that three switches are integrated into the meander line slot to control its resonance. By switching the switches ON and OFF, the resonance of the meander line slot can be well controlled. In this simulation, the presence of a

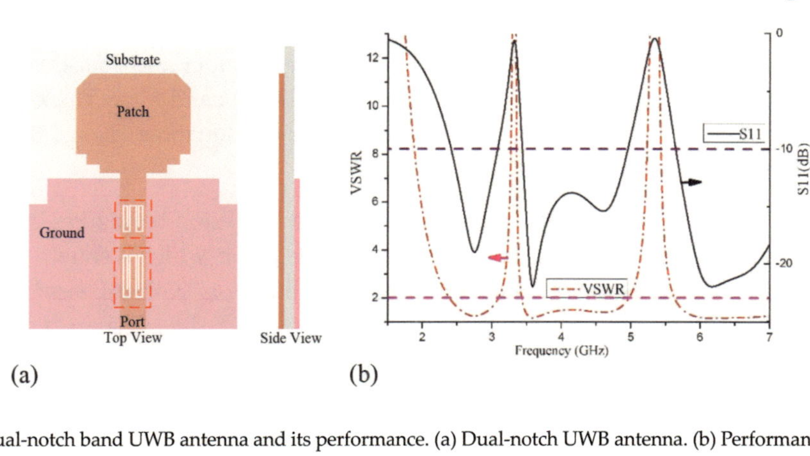

Figure 8. Dual-notch band UWB antenna and its performance. (a) Dual-notch UWB antenna. (b) Performance of the dual-notch UWB antenna.

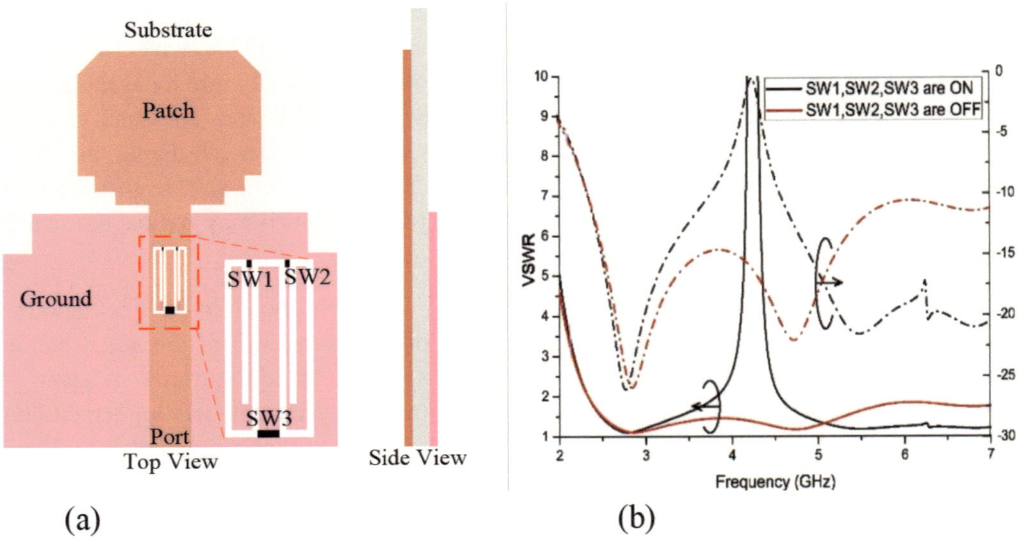

Figure 9. Reconfigurable UWB antenna. (a) Reconfigurable UWB antenna. (b) Behavior of the antenna.

metal bridge represents the ON state, while its absence represents the OFF state [26–30]. From **Figure 9(b)**, the antenna is a band-notched UWB antenna when all the switches are ON. A notch at 4.2 GHz is generated by the meander line slot. Moreover, the antenna is also a dual-band antenna. When all the switches are turned OFF, the antenna is changed to be a UWB antenna which has a bandwidth of 49.2%. Thus, the antenna can operate at least two modes for meeting the UWB and band-notched UWB communication requirements.

3. CR-UWB antenna

Based on the UWB antenna, band-notched UWB antenna, DMS stop-band filter design, switch techniques, and reconfigurable antenna design concept, CR-UWB antenna design will be introduced in detail. To obtain much more operation modes, three meander line slots are used for constructing the CR-UWB antenna. In this design, three meander line slots are etched in the microstrip transmission signal line, which is a DMS-based stop-band filter. Then, switches are integrated into these meander line slots to control their resonances. Here, the CR-UWB antenna design and analysis are given step by step.

To create a multiple mode CR-UWB antenna, a triple band DMS stop-band filter is designed. It is realized by etching three DMS cells in a 50-Ohm microstrip line, which is shown in **Figure 10**. It is observed that the three DMS cells are all meander line slots with different dimensions. Furthermore, these DMS cells can also be obtained by using spur slots, T-shaped slots, or their combinations. The performance of the triple stop-band filter is demonstrated in **Figure 11**. We can see that there are three stop bands operating at 3.3, 5.25, and 6.8 GHz, which can be used for preventing the potential interferences from 3.3 GHz WiMAX, 5.25 GHz WLAN, and 6.8 GHz RFID systems. Furthermore, the center frequencies of the three stop bands can be adjusted

http://dx.doi.org/10.5772/intechopen.69079

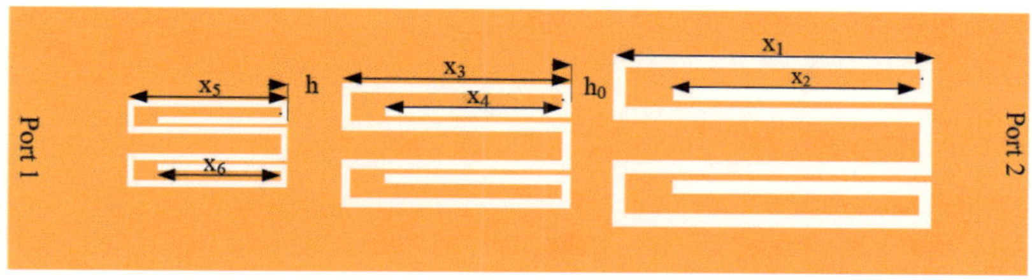

Figure 10. Geometry of the tri-band stop-band filter.

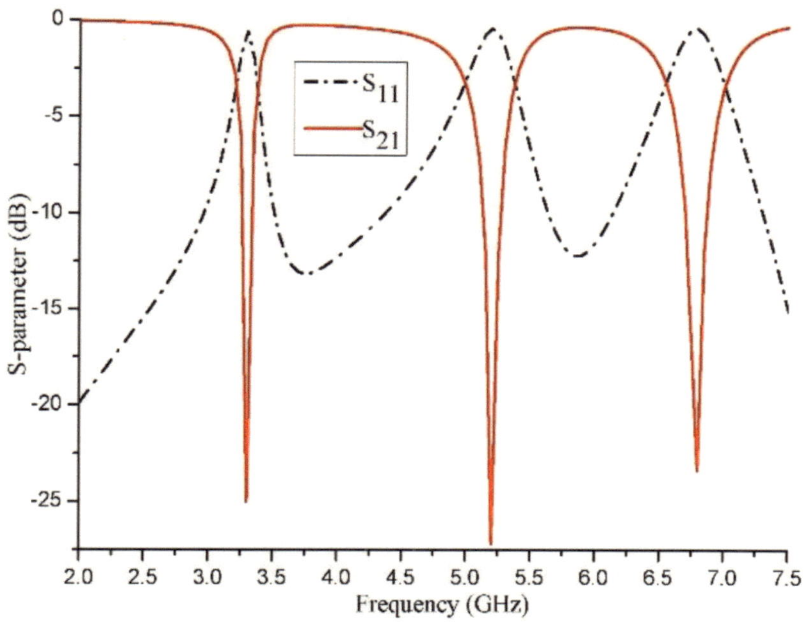

Figure 11. Performance of the tri-band stop-band filter.

by selecting the dimensions of these meander line slots, which are given in **Figure 12**. X_1, X_3, and X_5 control the length of the DMS resonators. With the increment of the X_1, the center frequency of the lowest stop-band shifts from high frequency to low frequency because the increased X_1 expands the resonance length of the right DMS cell. In this case, the center frequency of the middle stop band keeps constant. However, the center frequency of the highest stop band has a little shift when X_1 is 7.4 mm. Similarly, parameters X_3 and X_5 can well control the center frequencies of the middle and highest stop bands, respectively. However, X_3 and X_5 have slight effects on the other stop bands. Thus, the stop band can be adjusted by properly selecting the dimensions of these meander line slots.

The tri-band stop-band filter can be analyzed based on Butterworth low-pass filter theory which can calculate the circuit parameters of the DMS-based filters. The equivalent circuit model of the low-pass filter and the DMS cell is given in **Figure 13**. Here, g_0, g_1 and g_2 are normalized values, and g_0 is internal resistance, and g_1 and g_2 can be found from a **Table 1**.

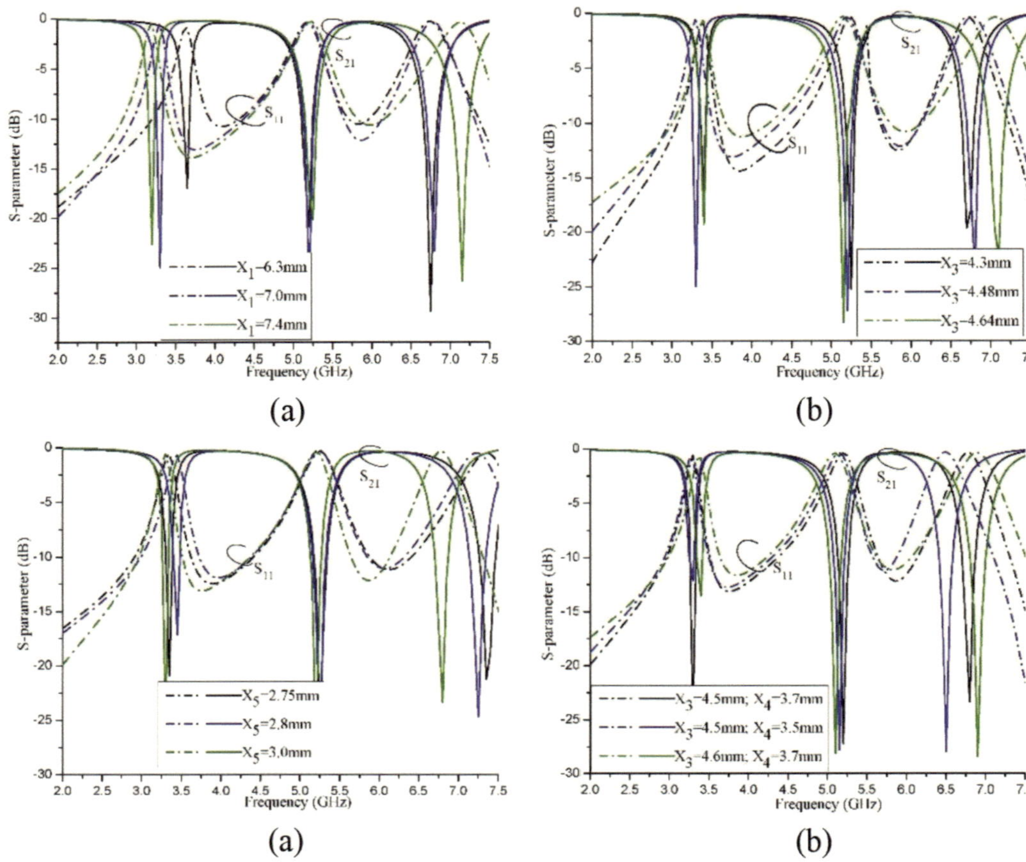

Figure 12. Parameter effects on the behavior of the tri-band stop-band filter. (a) X_1. (b) X_3. (a) X_5. (b) X_3 and X_4.

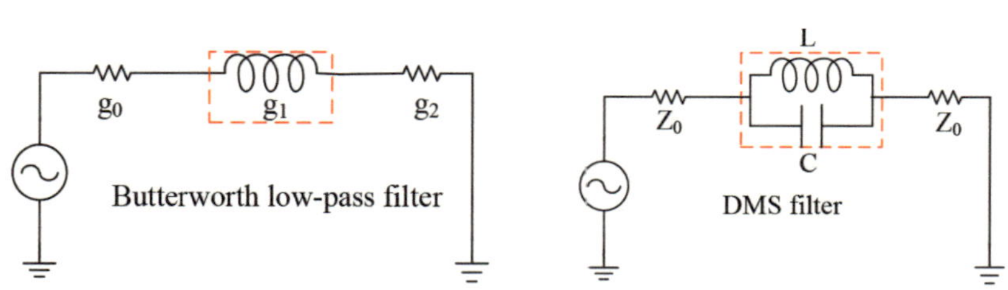

Figure 13. Equivalent circuit of the Butterworth low-pass filter and DMS filter.

Z_0 is characteristic impedance. Then, the reactance of the Butterworth low-pass filter and the DMS-based filter can be obtained Refs. [33, 34]

$$X_L = \omega' Z_0 g_1 \tag{1}$$

$$X_{LC} = \left[\omega_0 C\left(\frac{\omega_0}{\omega}\right) + \frac{\omega}{\omega_0}\right]^{-1}, \tag{2}$$

http://dx.doi.org/10.5772/intechopen.69079

Modes	SG1	SG2	SG3	Highest notch	Middle notch	Lowest notch
1	OFF	OFF	OFF	–	–	–
2	OFF	ON	OFF	–	4 GHz	–
3	OFF	OFF	ON	–	–	3.3 GHz
4	OFF	ON	ON	–	4 GHz	3.3 GHz
5	ON	OFF	OFF	5.5 GHz	–	–
6	ON	ON	OFF	6 GHz	4.5 GHz	–
7	ON	OFF	ON	5.5 GHz	–	3 GHz
8	ON	ON	ON	2.4 GHz	4 GHz	5.5 GHz

Table 1. Modes of the CR-UWB antenna.

where ω' is normalized angle frequency, ω_0 is the resonance frequency of the DMS cell and it is obtained from

$$\omega_0 = 1/\sqrt{LC} \tag{3}$$

Based on the circuit theory and the equivalent theory, we have

$$X_{LC}/_{\omega=\omega_C} = X_L/_{\omega'=1} \tag{4}$$

From Eqs. (1) and (4), we get Refs. [33, 34]

$$C = \left(\frac{\omega_C}{Z_0 g_1}\right) \frac{1}{\omega_0{}^2 - \omega_C{}^2}, \tag{5}$$

$$L = \frac{1}{4\pi^2 f_0{}^2 C} \tag{6}$$

where f_0 and f_c are the pole frequency and −3dB cut-off frequency. Since the tri-band stop-band filter is realized based on three cascaded DMS filters with different dimensions, the transmission network, including the capacitance and inductances, can be obtained

$$CP_i = -\frac{1}{2\pi f_{Ti} X_{(i+1),i}}, i = 1, 2, \tag{7}$$

$$LS_i = \frac{X_{ii} - X_{(i+1),i}}{2\pi f_{Ti}} + \frac{L_i}{(f_{Ti}/f_{0i})^2 - 1}, i = 1, 2, 3, 4, \tag{8}$$

where f_{0i} is the i-th pole frequency, f_{Ti} is the transmission poles between the pole frequencies, X is the imaginary part at f_{Ti}. Based on the theory above, the equivalent circuit model of the tri-band stop-band filter is obtained and is given in **Figure 14**. The values of the parameters are L_1 = 0.297 nH, C_1 = 7.839 pF, L_2 = 0.265 nH, C_2 = 3.541 pF, L_3 = 0.183 nH, C_3 = 3.003 pF, L_{s1} = 0.434 nH, L_{s2} = 0.01 nH, C_{p1} = 0.417pF, Ls3 = 0.852 nH, L_{s4} = −0.085 nH, C_{p2} = −0.08 pF.

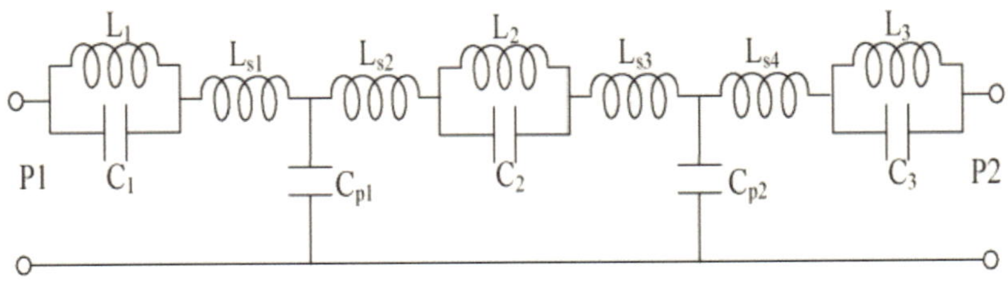

Figure 14. Equivalent circuit model of the tri-band stop-band filter.

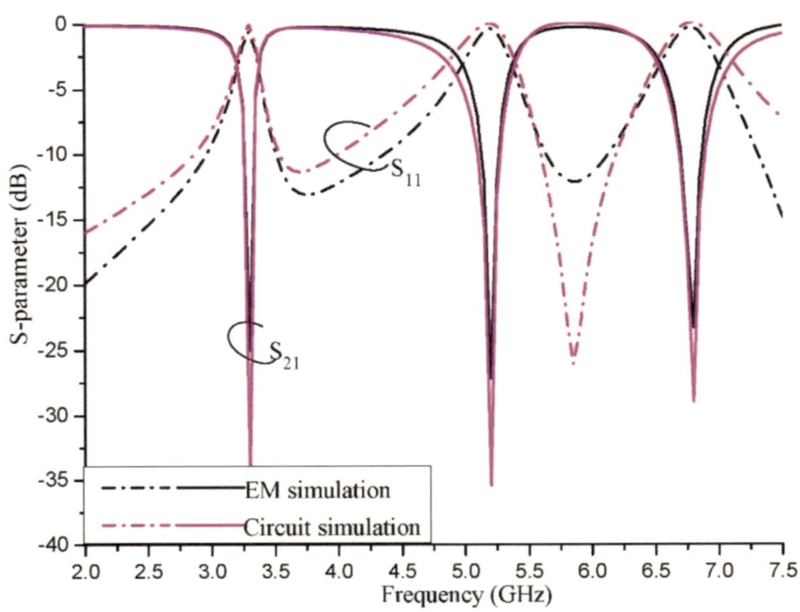

Figure 15. Comparisons of the tri-band stop-band filter.

The result of the circuit simulation is described in **Figure 15**. It is found that the circuit simulation agrees well with the EM simulation. Thus, the tri-band stop-band filter can be calculated based on circuit theory, which render it easy to understand and design. There is some fluctuation between the EM and circuit simulations, which can be corrected by carefully adjusting the values of the equivalent circuit.

Then, we use the designed tri-band stop-band filter to construct a triple band-notched UWB antenna. The stop-band filter is directly integrated into the feeding signal strip line to generate the desired three notches by properly choosing the dimensions of the DMS cells. The configuration of the band-notched UWB antenna with triple notches is shown in **Figure 16(a)**. It is observed that three DMS cells are sequentially etched on the transmission signal line and tapered structures are used to enhance the bandwidth of the UWB antenna. The impedance bandwidth of the tri-band band-notched UWB antenna is depicted in **Figure 16(b)**. The antenna has three notches at 2.8, 4, and 5.25 GHz to give resistance to the unwanted narrow-band signals. To further under the performance of the tri-band band-notched UWB antenna,

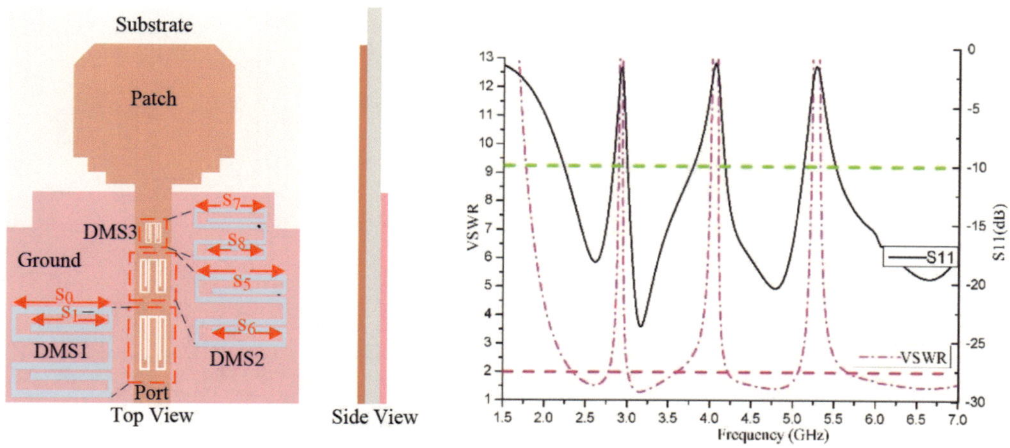

(a) (b)

Figure 16. Tri-band band-notched UWB antenna. (a) Geometry. (b) Impedance bandwidth of the antenna.

the parameters of the meander line slots are investigated. **Figure 17(a)** shows the design process of the tri-band band-notched UWB antenna. It is found that the antenna without any DMS cells is a UWB antenna which can provide a wide bandwidth. The antenna has a notch at 3.5 GHz when only DMS1 is integrated into the transmission signal line. As both DMS1 and DMS2 are incorporated into the feeding signal strip line, the antenna can generate two notches at 3.5 and 5.25 GHz, respectively. When the MDS1, DMS2, and DMS3 are used in the UWB antenna, the antenna has three notches at 2.8, 4, and 5.25 GHz. Since the size of the antenna and the feeding transmission signal line are limited, the coupling of these DMS cells might affect the resonance center frequencies. Thus, the lowest notch moves to low frequency. However, these notches are produced by independent DMS cells. **Figure 17(b)** shows the effects of S_0 on the impedance of the tri-band band-notched UWB antenna. It is observed that the lowest notch shifts from high frequency to low frequency when S_0 increases from 6.7 to 7.1 mm because the resonance length is expanded by the increased S_0. However, the center frequencies of the middle and the highest notches are also affected since the cascaded DMSs produce some couplings which may affect the circuit parameters. Similarly, the effects of the parameters S_3 and S_5 are discussed in **Figure 17(c)** and **(d)**, respectively. By properly selecting the S_3 and S_5, the center frequencies of the corresponding notches can be well adjusted to meet the practical engineering applications. Thus, these notches can be controlled by choosing the dimensions of the DMS1, DMS2, and DMS3.

To better understand the triple band-notched UWB antenna, the equivalent circuit model is extracted and given in **Figure 18(a)**. The parameter values are R_1 = 1.0 Ohm, L_1 = 58.804 nH, C_1 = 0.05 pF, R_2 = 59.0 Ohm, L_2 = 0.01 nH, C_2 = 1.509 pF, R_3 = 141.25 Ohm, L_3 = 0.5 nH and C_3 = 5.25 pF. The results are shown in **Figure 18(b)**. It is found that the circuit simulation is same as the EM simulation which helps to verify the effectiveness. There is some difference between the EM and circuit results which can be corrected by properly choosing the values of the circuit parameters.

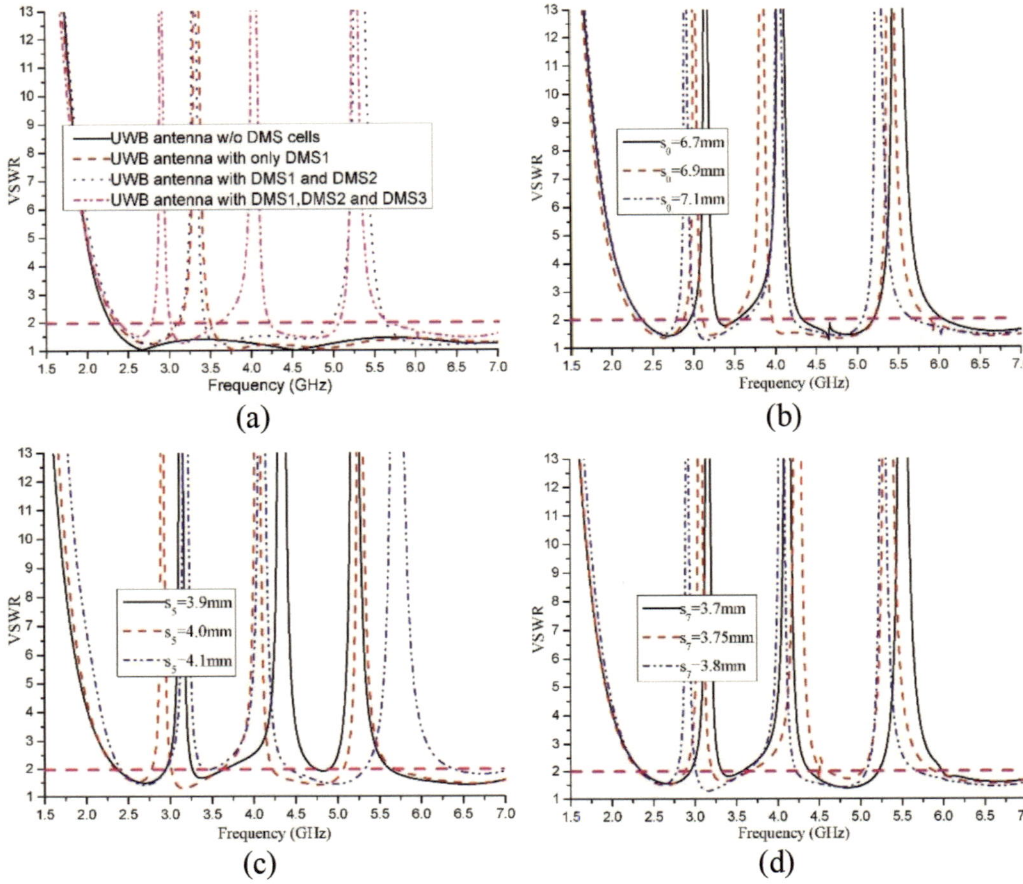

Figure 17. Tri-band band-notched UWB antenna. (a) Design process of the antenna. (b) S_0. (c) S_5. (d) S_7.

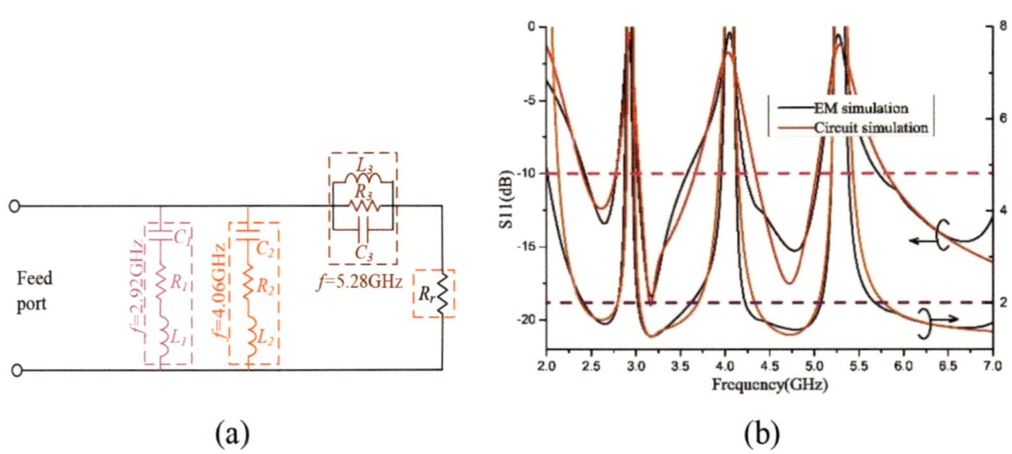

Figure 18. Equivalent circuit and the results of the triple band-notched UWB antenna. (a) Equivalent circuit model. (b) Impedance characteristics of the antenna.

Finally, nine switches are incorporated into the triple band-notched UWB antenna to carry out a CR-UWB antenna. The configuration of the CR-UWB antenna is shown in **Figure 19**. We can see that nine switches, namely, switch-1 (SW1), switch-2 (SW2), switch-3 (SW3), switch-4 (SW4), switch-5 (SW5), switch-6 (SW6), switch-7 (SW7), switch-8 (SW8), and switch-9 (SW9), are used for realizing the CR-UWB antenna. In the design, the SW1, SW2 and SW3 are named as switch group 1 (SG1), and the SW4, SW5 and SW6 are denoted as switch group 2 (SG2), and the SW7, SW8 and SW9 are formed to be switch group 3 (SG3). All the switches in each group are turned ON or turned OFF instantaneously.

The operating modes of the CR-UWB antenna are shown in **Figure 20**. It is found that there are 8 operating modes by using the three group switches. In the simulation, the presence of a metal

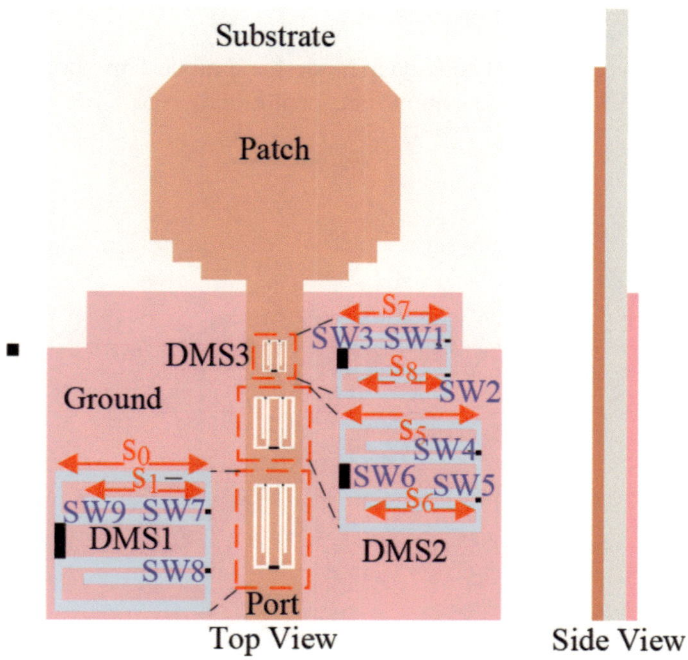

Figure 19. Configuration of the CR-UWB antenna.

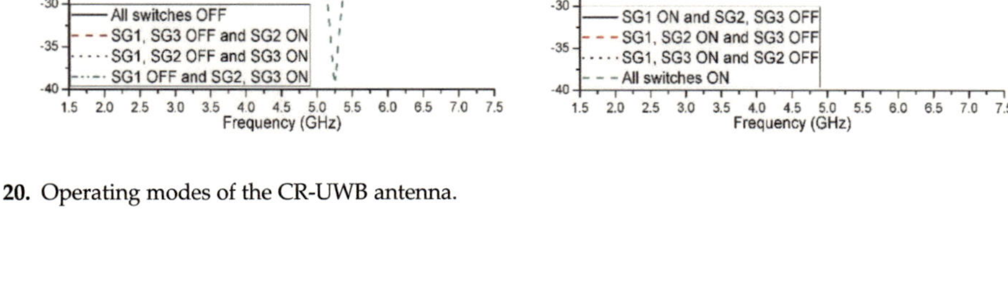

Figure 20. Operating modes of the CR-UWB antenna.

bridge represents the ON state, while its absence represent OFF state. The operating modes are given in **Table 1**. When the CR-UWB antenna works in mode 1, it is a UWB antenna which can be used in underlay mode in a very low power. In this case, the IR-UWB technology can be used to transmit and receive desired signals. Furthermore, it can also be used as a sensing antenna in CR communication systems. As the CR-UWB antenna operates at mode 2 to mode 8, it is a UWB antenna with different notches. In these cases, the CR-UWB system can be implemented by using OFDM-UWB technology to switch ON/OFF the different carries to make the antenna to prevent the unwanted narrowband signals. Thus, the designed antenna can be used for various CR-UWB communication systems to sense and to prevent interferences. It can also switch between the overlay and underlay modes by change the antenna operating modes. In addition, the CR-UWB antenna can be used for UWB, band-notched UWB and multiband communication systems.

The radiation patterns of CR-UWB antenna in mode 4 and mode 5 are shown in **Figure 21**. It is found that the CR-UWB antenna has omnidirectional radiation patterns in the H-plane and it

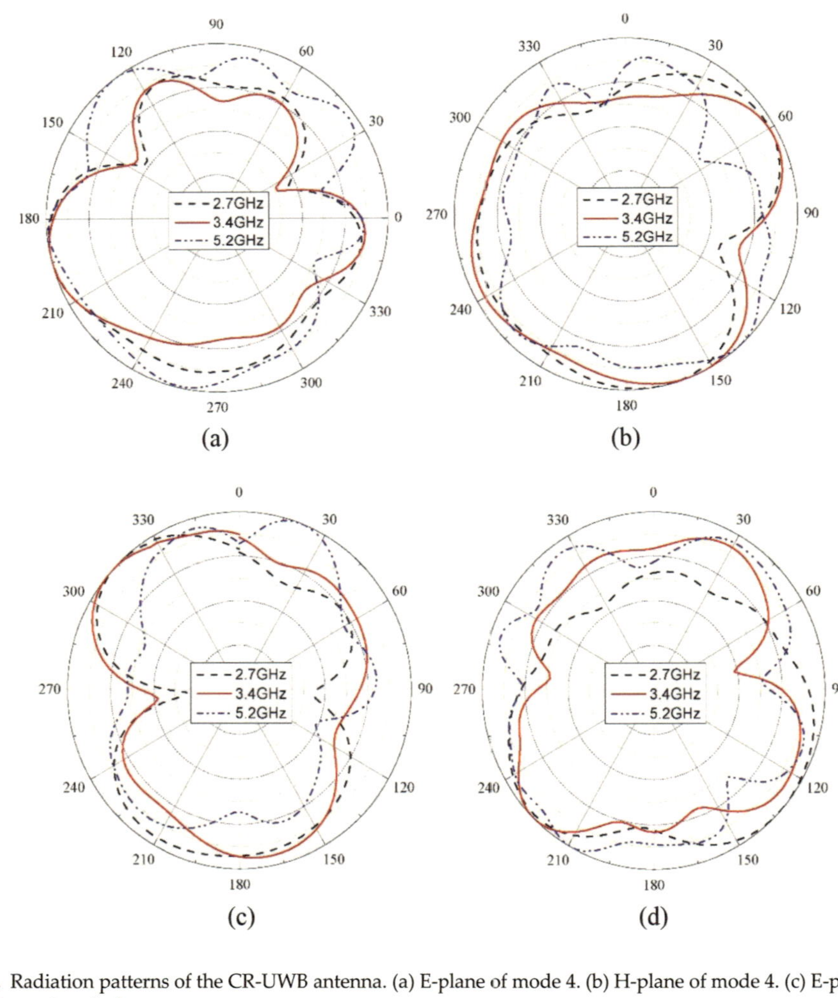

Figure 21. Radiation patterns of the CR-UWB antenna. (a) E-plane of mode 4. (b) H-plane of mode 4. (c) E-plane of mode 5. (d) H-plane of mode 5.

can provide eight-like radiation patterns in its E-plane, which render the CR-UWB antenna suitable for multiple mode communication requirements. The radiation patterns in other modes are similar as the mode 4 and mode 5.

4. Conclusion

In this chapter, UWB antennas for CR communication have been reviewed. The DMS-based stop-band filter, UWB antenna, band-notched UWB antenna and reconfigurable UWB antenna have been discussed to construct a CR-UWB antenna. The CR-UWB antenna has been realized by integrating desired DMS-based stop-band filters and radio frequency switches into a UWB antenna. The DMS-based filters were used to create the notches to filter out the unwanted narrowband interference signals, while the switches control the reconfigurable modes of the antenna. The antenna is designed step by step and it is analyzed in detail. The results showed that the CR-UWB antenna can be used to switch between different modes and it can be used as UWB antenna, band-notched UWB antenna and multiband communication systems.

Acknowledgements

This work was also partially supported by the National Key Research and Development Program of China-Government Corporation Special Program (2016YFE0111100), National Science Foundation of China (61571149), the Science and Technology Innovative Talents Foundation of Harbin (2016RAXXJ044), Projects for the Selected Returned Overseas Chinese Scholars of Heilongjiang Province and MOHRSS of China, and the Foundational Research Funds for the Central Universities (HEUCF160815, HEUCFD1433).

Author details

Yingsong Li* and Yanyan Wang

*Address all correspondence to: liyingsong@ieee.org

Harbin Engineering University, Harbin, China

References

[1] Li YS, Yang XD, Liu CY Jiang T. Compact CPW-fed ultra-wideband antenna with dual band-notched characteristics. Electronics Letters. 2010;**46**(14):967–968

[2] Liang J, Chiau CC, Chen X, Parini CG. Study of a printed circular disc monopole antenna for UWB systems. IEEE Transactions on Antennas and Propagation.2005;**53**(11):3500–3504

[3] Choi SH, Park JK, Kim SK, Park JY. A new ultra-wide band antenna for UWB applications. Microwave and Optical Technology Letters. 2005;**53**(11):3500–3504

[4] Li YS, Yang X, Li Q, Yang Q. Compact coplanar waveguide fed ultra wideband antenna with a notch band characteristic. AEU-International Journal of Electronics and Communications. 2011;**65**(11):961–966

[5] First Report and Order. Revision of Part 15 of the Commission's Rule Regarding Ultra-Wideband Transmission System FCC 02-48. Federal Communications Commission; USA, 2012

[6] Zhao L, Haimovich AM. Performance of ultra-wideband communications in the presence of interference. IEEE Journal on Selected Areas in Communications. 2002;**20**(9):1684–1691

[7] Oppermann I, Stoica L, Rabbachin A, Shelby Z, Haapola J. UWB wireless sensor networks: UWEN – A practical example. IEEE Communications Magazine. 2004;**42**(12):S27-S32

[8] Moallemizadeh A, Hassani HR, Mohammad ali nezhad S. Wide bandwidth and small size LPDA antenna. 6th European Conference on Antennas and Propagation (EUCAP), Prague, 2012;1–3

[9] Dehdasht-Heydari R, Hassani HR, Mallahzadeh AR. Quad ridged horn antenna for UWB applications. Progress in Electromagnetics Research. 2008;**79**:23–38

[10] Chen ZN, See TSP, Qing X. Small printed ultrawideband antenna with reduced ground plane effect. IEEE Transactions on Antennas and Propagation. 2007;**55**(2):383–388

[11] Ojaroudi M, Ghobadi C, Nourinia J. Small square monopole antenna with inverted t-shaped notch in the ground plane for UWB application. IEEE Antennas and Wireless Propagation Letters. 2009;**8**:728–731

[12] Li Y, Li W, Yu W. A switchable UWB slot antenna using SIS-HSIR and SIS-SIR for multi-mode wireless communications applications. Applied Computational Electromagnetics Society Journal. 2012;**27**(4):340–351

[13] Li X, Hagness SC, Choi MK, van der Weide DW. Numerical and experimental investigation of an ultrawideband ridged pyramidal horn antenna with curved launching plane for pulse radiation. IEEE Antennas and Wireless Propagation Letters. 2003**2**(1):259–262

[14] Ammann MJ, Chen ZN. Wideband monopole antennas for multi-band wireless systems. IEEE Antennas and Propagation Magazine. 2003;**45**:146–150

[15] Wang HN, Li Y. Bandwidth enhancement of a wide slot UWB antenna with a notch band characteristic. IEEE 3rd International Conference on Communication Software and Networks (ICCSN), Xi'an, China. 2011;365–368

[16] Li YS, Yang XD, Liu CY, et al. Compact CPW-fed ultra-wideband antenna with band-notched characteristics. Electronics Letters. 2010;**46**(23):1533–1534

[17] Chu Q-X, Yang Y-Y. A compact ultrawideband antenna with 3.4/5.5 GHz dual band-notched characteristics. IEEE Transactions on Antennas and Propagation. 2008;**56**(12): 3637–3644

[18] Barbarino S, Consoli F. UWB circular slot antenna provided with an inverted-L notch filter for the 5 GHz WLAN band. Progress in Electromagnetics Research. 2010;**104**:1–13

[19] Zhu X-F, Su D-L, Symmetric E-shaped slot for UWB antenna with band-notched characteristic. Microwave and Optical Technology Letters. 2010;**52**(7):1594–1597

[20] Li Y, Li W, Yu W. A compact reconfigurable antenna using SIRs and switches for ultra wideband and multi-band wireless communication applications. Applied Computational Electromagnetics Society Journal. 2013;**28**(5):427–440

[21] Li Y, Li W, Ye Q. A compact circular slot UWB antenna with multimode reconfigurable band-notched characteristics using resonator and switch techniques. Microwave and Optical Technology Letters. 2014;**56**(3):570–574

[22] Li Y, Li W, Yu W. A CPW-fed circular slot UWB antenna with WLAN band and X-band filtering characteristics using hybrid resonators. Microwave and Optical Technology Letters. 2014;**56**(4):925–929

[23] Kim J, Cho CS, Lee JW. 5.2 GHz notched ultra-wideband antenna using slot-type SRR. Electronics Letters. 2006;**42**(6):315–316

[24] Li Y, Li W, Ye Q. A compact UWB antenna with dual band-notch characteristics using nested split ring resonator and stepped impedance resonator. Microwave and Optical Technology Letters. 2013;**55**(12):2827–2830

[25] Li Y, Li W, Mittra R. A CPW-fed wide-slot antenna with reconfigurable notch bands for UWB and multi-band communication applications. Microwave and Optical Technology Letters. 2013;**55**(11):2777–2782

[26] Li Y, Li W, Ye Q. A reconfigurable triple notch band antenna integrated with defected microstrip structure band-stop filter for ultra-wide band cognitive radio applications. International Journal of Antennas and Propagation. 2013;**2013:**1–13

[27] Tawk Y, Christodoulou CG. A new reconfigurable antenna design for cognitive radio. IEEE Antennas and Wireless Propagation Letters. 2009;**8**:1378–1381

[28] Tawk Y, Costantine J, Avery K, Christodoulou CG. Implementation of a cognitive radio front-end using rotatable controlled reconfigurable antennas. IEEE Transactions on Antennas and Propagation. 2011;**59**(5):1773–1778

[29] Li Y, Li W, Mittra R. A cognitive radio antenna integrated with narrow/ultra-wide band antenna and switches. IEICE Electronics Express. 2012;**9**(15):1273–1283

[30] Li Y, Li W, Ye Q. A reconfigurable wide slot antenna integrated with SIRs for UWB/multi-band communication applications. Microwave and Optical Technology Letters. 2013;**55**(1): 52–55

[31] AI-Husseini M, Safatly L, Ramadan A, et al. Reconfigurable filter antennas for pulse adaptation in UWB cognitive radio systems. Progress in Electromagnetics Research. 2012;**37**:327–342

[32] Safatly L, Bkassiny M, AI-Husseini M, et al. Cognitive radio transceivers: RF, spectrum sensing, and learning algorithms review. International Journal of Antennas and Propagation. 2014;**2014**:2014

[33] Zhang S, Xiao J-K, Wang Z-H, and Li Y. Novel low pass filters using a defected microstrip structure. Microwave Journal. 2006;**49**(9):118–128

[34] Wang Y. The study design of antenna integrating with stop-band filter [thesis]. Master Thesis of Harbin Engineering University; Harbin, China. 2016